# SOMEWHERE INSIDE

WILLIAM MORROW

*An Imprint of* HarperCollins*Publishers*

# SOMEWHERE

## inside

ONE SISTER'S CAPTIVITY
IN NORTH KOREA
AND THE OTHER'S FIGHT
TO BRING HER HOME

LAURA LING AND LISA LING

All insert photographs, unless otherwise indicated, are from the Ling family collection. Page 6, top and bottom: photographs by Eric Powell; page 10, top: photograph by Bob Toy; page 11, top: photograph by Rory White; page 11, bottom, and page 12, top: photographs courtesy of Sara Mibo Sohn; page 12, bottom: photograph by Hector Amezcua; page 13, top: reprinted with permission of the Associated Press / KRT TV via APTN; page 13, bottom: photograph by Morgan Wandell; page 15, top: photograph by Robyn Beck / Getty Images; page 15, bottom: photograph by Charles Clayton.

HarperCollins books may be purchased for educational, business, or sales promotional use. For information please write: Special Markets Department, HarperCollins Publishers, 10 East 53rd Street, New York, NY 10022.

FIRST EDITION

*Designed by Jamie Lynn Kerner*

Library of Congress Cataloging-in-Publication Data

Ling, Laura, 1976–
    Somewhere inside : one sister's captivity in North Korea and the other's fight to bring her home / Laura Ling and Lisa Ling. — 1st ed.
        p.   cm.
    ISBN 978-0-06-200067-5
    1. Ling, Laura, 1976– —Captivity, 2009. 2. Ling, Lisa. 3. Hostages—Korea (North) 4. Journalists—Korea (North) 5. Americans—Korea (North) 6. Korea (North)—Politics and government—1994– 7. Journalists—United States—Biography. 8. Clinton, Bill, 1946– I. Ling, Lisa. II. Title.
    PN4841.L56A3   2010
    365'.45092—dc22
    [B]
                                                          2010012124

10  11  12  13  14   ov/RRD   10 9 8 7 6 5 4 3 2 1

*Dedicated in the hope that all people*
*will one day experience freedom*

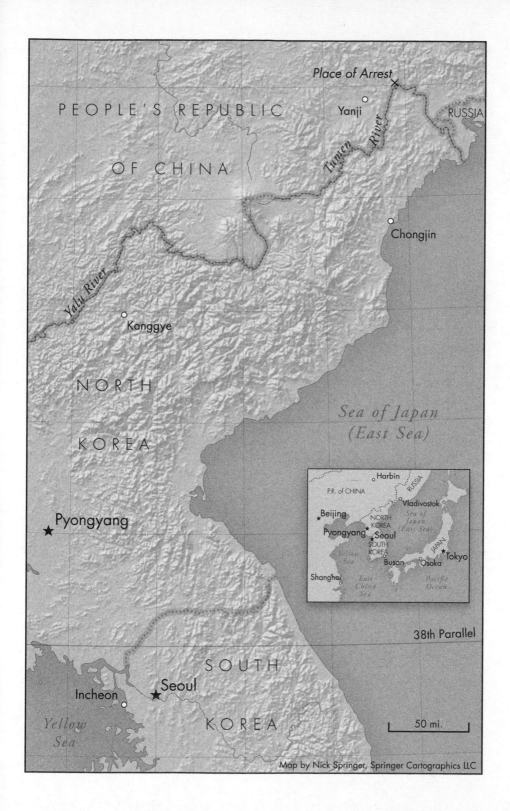

PEOPLE'S REPUBLIC

OF CHINA

Place of Arrest ✕

Yanji

RUSSIA

*Tumen River*

Chongjin

*Yalu River*

Kanggye

NORTH

KOREA

*Sea of Japan
(East Sea)*

Pyongyang ★

Harbin

P.R. of CHINA

RUSSIA

Beijing ★

Vladivostok

NORTH
KOREA

*Sea of
Japan
(East Sea)*

Pyongyang ★

★ Seoul

SOUTH
KOREA

JAPAN

★ Tokyo

Shanghai

Busan ★

Osaka

*Yellow
Sea*

*East
China
Sea*

*Pacific
Ocean*

38th Parallel

SOUTH

Seoul ★

Incheon

KOREA

*Yellow
Sea*

50 mi.

# contents

# authors' note

THIS IS A WORK of nonfiction that is primarily about our experiences in 2009 when Laura was seized by North Korean soldiers and held in that country for nearly five months. Laura has given names to some of the people she encountered in North Korea, but she never actually knew their real names and referred to them, if she did, as "sir" or "ma'am." She has also changed the names of North Koreans whom she interviewed before her apprehension to protect their identities. Lisa has changed the names of two people she worked with, at their request, to protect their anonymity.

# preface

WE WERE JUST FOUR and seven years old when our immigrant parents divorced. Few other parents at the time were separating in our all-American suburban community, and that filled us with insecurities and confusion. At least we had each other and could be each other's protector and close confidante. It is impossible to measure the bond that formed between us.

Our grandmother lived with us during our parents' divorce. She was a lady of strong Christian faith and character, and she encouraged us to be determined women and to stand up for people who didn't have a voice. We took her words and lessons to heart.

As kids, we fantasized about escaping to distant lands. We played a game that involved a spaceship that could transport us from place to place, where we could embark on amazing adventures, battling villains and coming to the aid of those in need.

As adults, we found that through journalism, we could open people's eyes to what was happening in the real world, just as Grandma had encouraged us to do. Between the two of us, we've spent more than twenty-five years traveling the globe.

We've seen things during our journeys that have moved us, from an Indian sex worker who has devoted her life to saving girls on the street, to ex-gang members in Los Angeles trying to bring positive change to their communities, to people rescuing children from child-trafficking rings in Ghana. We've also encountered things that have scarred us, from women violently gang-raped in the Democratic Republic of the Congo, to people forced into slavery in the jungles of Brazil, to whole communities ravaged by toxic pollutants in China.

These experiences have filled us with a desire to tell the world about the people we've met and the things we have witnessed. We have been driven by a passion to try to be the eyes and ears for people who wish to explore unfamiliar cultures.

When, in March 2009, one of us got into trouble while reporting a story about the thousands of people being trafficked from North Korea into China, the other one jumped into action to try to help. Our bond as sisters and best friends got us through this horrifying time, even though we were thousands of miles apart. We drew strength from somewhere inside.

During this period of darkness, we experienced rays of light. They came in the form of unexpected relationships that evolved even in this time of crisis. One of us developed a better understanding of her captors and they of her. The other was helped by loads of people, many of whom she'd never met, who showed up to offer support.

Throughout it all, we were able to experience what happens when human beings get a chance to interact face-to-face, eye-to-eye, even if their countries are "enemies."

This is our story.

# CHAPTER ONE

# somewhere inside north korea

*Dearest Lisa,*

*Please do not share this letter with Mom or Dad, as I do not want them to worry. I am trying so hard to be strong, but it gets harder and harder every day. It is so difficult to get through each day. I miss you all so much it hurts. I want my big sister.*

*As I'm sure you know, I am in the worst possible situation. . . .*

## ∽≺ LAURA

WE ARRIVED IN YANJI, China, on March 13, 2009. The mountainous region that borders Russia and North Korea is one of China's coldest. As our team walked out of the airport, I clenched my fists tightly and hid my face in my woolen scarf to protect me against the bone-chilling, cloud-covered night. Over the past decade, I have made more than half a dozen trips to China—it's where my father and his forefathers are from, and it's always been one of the most fascinating places to work as a journalist. I'd reported from different parts of the vast country, but this was my first time in the northeast, where a large portion of the population is of Korean ancestry. The project we were working on had as much to do with something happening in neighboring North Korea as it did with this part of China, and being in Yanji, I could immediately sense a connection between the Korean and Chinese cultures. Signs are written in both Korean and Chinese characters; most of the restaurants serve Korean food. It would be easy for someone of Korean descent to blend in, without knowing a single word of Chinese.

Our small team consisted of producer/cameraman Mitchell Koss, coproducer/translator Euna Lee, and myself. We had traveled to the area to investigate a controversial issue to which neither the North Korean nor the Chinese government wants any attention drawn. Millions of citizens of North Korea, one of the most isolated, repressive countries in the world, suffer from dire poverty and brutal conditions, and some of them take the risk of fleeing, or defecting, from their homeland by crossing the border into neighboring China. But once in China, they end up facing a different kind of degradation.

China classifies these defectors not as refugees, but as illegal immigrants so rather than finding safe haven across the border, most of them end up in hiding, living underground in fear of being arrested by Chinese authorities. Those who are caught and repatriated back to North Korea could be sent to one of the country's notorious gulags, where they face torture or possibly execution.

Most of these defectors are North Korean women who are preyed on by traffickers and pimps. These women escape from their country to find food; some are promised jobs in the restaurant or manufacturing industries. But they soon find out that a different, dark fate awaits them. Many end up being sold into marriages or forced into China's booming sex industry. I wanted to open people's eyes to the stories of these despairing women who are living in a horrible, bleak limbo with no protection or rights.

On our first night in Yanji, our three-person team arranged to meet up with the man we'd hired to be our guide. He was referred to us by a Seoul-based missionary, Pastor Chun Ki-Won, who has become a kind of legend in the area for helping North Korean defectors find passage to South Korea through an underground network. Our guide had worked with Chun as well as other foreign journalists in the past. He was also a kind of smuggler himself, with deep connections in North Korea. He claimed to have a clandestine operation in North Korea that loaned out Chinese cell phones to North Koreans and, for a fee, let them call relatives or friends in China or South Korea. Telephone use is strictly controlled in North Korea, and making calls outside of the country without permission is almost impossible and dangerous.

We met our guide, a Korean-Chinese man who appeared to be in his late thirties, at our hotel to discuss our plans. His reserved demeanor and deadpan expression made him a hard read. We were hoping he could introduce us to some defectors and take us to the border area where North Koreans make their way to China. He said he could make the arrangements, but emphasized the risky nature of our investigation. We knew we would have to be cautious and discreet so we didn't put any defectors at risk of deportation.

Before leaving for China, our team had decided to forgo applying for journalist visas. Normally, foreign journalists working in China are required to have a special visa and must also work with a Chinese media entity. But because of the nature of our story and the sensitivity with which the Chinese government regards the issue of North Korean defectors, we decided to enter the country as tourists. We didn't want to draw attention to the people we were interviewing, so as not to endanger them or ourselves. We would be careful to conceal the identity of defectors when we filmed them, focusing on body parts or the backs of heads rather than faces or easily identifiable features.

-❦- **LISA**

I HADN'T BEEN PARTICULARLY worried about Laura's assignment to the Chinese–North Korean border. A month earlier she had been in Juárez, Mexico, a city that had a higher death rate than Baghdad. The *Los Angeles Times* regularly carried headline stories about law enforcement officers and journalists being attacked by narco-traffickers. Every day Laura was there, I was struck by episodes of paralytic concern. She and her producer were shadowing Mexican homicide reporters who were chasing death. The documentary that aired showed one gruesome crime scene after another—from corpses left in a trash-filled ravine to mutilated bodies riddled with dozens of bullet holes. Needless to say, our family breathed a major sigh of relief when Laura was finally back from that assignment. She was so preoccupied with getting the Mexico show on the air that she never even told me she was going to Asia several weeks later. It was almost an afterthought when she mentioned that soon she would be leaving for another trip.

"What are you doing?" I pressed. "You just came back. I thought you were going to stop traveling so much."

"I know, Li," she replied. "Don't worry. Everything is already set up."

Laura and her team were headed first to Seoul and then to China's border with North Korea to meet up with contacts and do some prearranged interviews. The trip was supposed to last a week and a half. My husband, Paul, even made a dinner reservation at a new barbecue restaurant for the Friday of Laura's return. Still, none of us, including our parents and Laura's husband, Iain, were eager for her to go. She had just wrapped up an extensive assignment, and we felt she had been working too hard recently. But arguing with Laura was pointless. She had always put a great deal of pressure on herself. She never stopped working.

Whenever we were together, she constantly checked her BlackBerry no matter what was going on around her. I'm a self-professed BlackBerry addict too, but Laura put me to shame. I found myself constantly frustrated by her lack of attention to anything but work. A few times I noticed it taking a toll on Iain. More than once I tried to scare her by telling her she better start paying more attention to her husband or he might find someone else who would.

## ∿৵ LAURA

FOR THIS ASSIGNMENT ALONG the Chinese–North Korean border, one of my two colleagues was Mitch Koss, someone with whom I'd been working closely for the past several years. Mitch had been a mentor to me, a driving force in my decision to pursue journalism. I also considered him a member of my extended family. He'd worked with my sister, Lisa, when she was just starting her journalism career. After Lisa left Channel One News, where she and Mitch had worked together for five years, he approached me to help him with an assignment as a researcher. I jumped at the opportunity.

Over the years, Mitch and I worked on more than three dozen stories spanning the globe, including a visit in the summer of 2002 to North Korea's capital, Pyongyang, where we, along with the

Korean-American tour group to which we were assigned, were taken on a highly monitored tour of the capital's most impressive monuments and sights.

Then in 2005 I was hired by Current TV, former Vice President Al Gore's cable network, to develop its journalism department. Mitch was also brought on board by Current to advise other young journalists. Each week, our unit produced a half-hour investigative documentary program called *Vanguard*. In addition to my role as manager of the sixteen-person team, I was also one of the on-air correspondents, reporting from various locations around the world. In the past year, I had covered China's restive Muslim population, life on parole in America, and Mexico's drug war. Now I was here in China's frigid northeast reporting on the trafficking of North Korean women.

My other colleague, Euna, was an editor in our journalism department. Because of her fluency in Korean, she was working on the project as a translator as well as a coproducer. Euna is a Korean American and I knew this made her particularly devoted to the assignment. She had been in communication with Pastor Chun in advance of our trip and, with his help, made most of our filming arrangements.

On a hazy, overcast morning, one day after our arrival in Yanji, our guide drove us two hours away to a logging town along the Chinese–North Korean border. We arrived at a small, dusty village where we met with Mrs. Ahn, a woman who appeared to be in her early fifties. She had fled North Korea in the late nineties at the height of a devastating famine. Estimates vary, but it's believed that anywhere from hundreds of thousands to perhaps two million people died as a result of the famine. Conditions were so dire during that time that many North Koreans attempted to escape to China, where they heard they could get white rice, which had become virtually nonexistent in North Korea. Defectors, including Mrs. Ahn, bribed

North Korean border guards to let them cross the river into China. Some hired so-called brokers to guide them across the treacherous waters. But once in China, many found themselves lost, with no way to make a living. The brokers, taking advantage of their vulnerable state, ended up selling these desperate women to Chinese men as wives.

The selling of women as brides is becoming increasingly rampant throughout China. In 1979 the Chinese government, in reaction to its exploding population, began limiting to one the number of children Chinese couples could have. The policy became known as the One Child Policy. What the government did not anticipate was that so many couples would want that one child to be male. As a result, tens of thousands of Chinese baby girls were aborted or abandoned, and today the country has tens of millions more males than females. Already men are having a difficult time finding wives, and women are being trafficked from other parts of the world, including North Korea, to fill this role. The women are sold off like animals to Chinese men, many of whom live in China's impoverished countryside. While these women may receive more sustenance living as purchased brides, they exist without residency certification or identification cards, which means that at any point they can be arrested and sent back to North Korea, where they face certain punishment.

Not only is the reality grim for these women defectors, but the children they bear to Chinese husbands also suffer. The Chinese government does not view the marriages of North Korean defectors to Chinese men as legitimate and therefore does not recognize these children as citizens. If the mothers are repatriated to North Korea or resold to other men, as sometimes happens, the fathers often end up abandoning the children. Some of these children are cast away because their fathers are too old and disabled to care for them. With no identification cards, they are unable to attend school, and they are denied health care; they must live in the shadows as stateless children. At a clandestine foster home run by a missionary in Pastor Chun's network, we met with a half-dozen foster children between the ages of six and ten who were being given clothes, a warm, clean place to live, and an education. It was hard to realize that without the help of Chun's group, these young souls might be roaming the streets without any parents or a government to provide for them. They would be lost and without identities.

Although conditions have improved in North Korea since the famine of the 1990s, a new generation of defectors is fleeing the country because the situation remains bleak and hunger is widespread. North Korea has maintained its overwhelming control over its citizens in part because of a propaganda machine that over the years has caused its people to believe that the rest of the world has been suffering even more than North Korea has. But little by little, information seems to be seeping into to the country.

The demilitarized zone, or DMZ, separating the two Koreas is the most heavily fortified border in the world. Soldiers on either side patrol their respective area along the thirty-eighth parallel, where in 1953 U.S. administrators divided the peninsula, three years after the start of the Korean War. The war was suspended by an armistice, but it never officially ended, meaning that the two sides are technically still at war. Because of the DMZ's impenetrable barrier, where on one side thousands of U.S. troops support South Korean forces, and where nearly a million North Korean soldiers are stationed on the opposing side, it has been relatively easy for the North to keep information from high-tech South Korea from flowing into its country.

China, on the other hand, is North Korea's closest ally. The border between the two countries is extremely porous. In many areas there are no fences or actual barriers, only a narrow river, separating the two countries. As a result, a thriving black market has emerged in North Korea as Korean-Chinese businesspeople take advantage of the North's isolation. Not only do products from China get across the border, so does knowledge about China's economic prosperity. North Korea, the so-called Hermit Kingdom, is finding it harder and harder to keep information about the rest of the world from coming across its border.

One afternoon we took a taxi from our hotel in Yanji to a nearby location where we had arranged to meet with a young woman who had fled North Korea the previous year. Ji-Yong was in her early twenties and had a round baby face. She looked as if she was playing dress up in her black go-go boots, long, thick false eyelashes, and electric blue eye shadow. We picked her up and drove her back to our hotel.

While Ji-Yong was able to eat three meals a day in her village in North Korea, the portions were very small. Many North Koreans only receive meat on very special holidays, which occur roughly three times a year. Ji-Yong, like an increasing number of young North Korean women, was told by a broker that he could find her a job making good money working with computers. Unable to swim, and in the dark of night, these girls brave the cold, rushing water of the bordering river to reach the other side, where the promise of opportunity awaits them. Some perish along the way. While the broker did arrange for Ji-Yong to work with computers, it wasn't the office job she had envisioned. She was placed in the online sex industry, video chatting with clients and undressing for them online.

Many women like Ji-Yong are filling the ranks of China's growing prostitution and Internet sex world. They must pay back large sums in order to win their freedom, an almost impossible task given their paltry wages. Some are beaten and confined in their working quarters. Others are afraid to leave for fear of being arrested and deported.

I WAS PROUD OF THE STORY Laura would be reporting. We had both covered a number of stories about sexual trafficking throughout our careers and felt strongly about the issue. When our mother was a child growing up in Taiwan, she had seen desperate women having to sell their bodies to survive. Her stories both enraged and touched us as women, and as young journalists we sought to raise awareness about the global sexual exploitation of women whenever possible.

But in recent months, I was starting to get very concerned that Laura was overworking herself. Her self-imposed pressure was unrelenting. It hurt me to see how much her work was bleeding into her personal life, even to the point where it started affecting her health. She had literally made herself sick from taking on so much.

Our family was most concerned about the recurring ulcers she had been dealing with for more than a year. I can't recall how many times Laura would call from another country to tell me about her stomach ailments, which seemed to be made worse by severe foreign environments. She had been on medication for more than a year, and her last endoscopy indicated that though her original ulcer had shrunk, a new one had formed. I was with her during that procedure and became deeply saddened because she and Iain had been seriously thinking of starting a family, and she didn't want to do that while she was on ulcer medication. My husband, Paul, is a physician, and he was also becoming concerned about her health. He remarked a number of times that Laura "really needs to lower her stress levels."

I know I was annoying Laura by constantly urging her to slow down. She would often shoot back with "You're one to talk. You travel as much as I do, if not more." While this was true, I wasn't managing a department simultaneously, and I rarely got sick on the road. Plus, she was my little sister, and looking out for her was the role I'd always played.

A

FTER THREE DAYS OF FILMING in Yanji and its outlying villages, the three of us, along with our guide, went to a café to discuss our next day's filming plans. It was 9:00 P.M. and we'd just finished interviewing a defector in a small town close to the North Korean and Russian border. Signs for restaurants and shops were written in all three languages: Chinese, Russian, and Korean. We passed a row of brothels disguised as massage parlors and could see groups of young women waiting inside rooms that were dimly lit with red lightbulbs.

I was exhausted. When we landed in the region a week earlier, we hadn't allowed any time to get over the jet lag of a sixteen-hour flight and the sixteen-hour time difference. But it wasn't the lack of sleep that was getting to me. I was feeling emotionally drained from hearing the harrowing life stories of the defectors we'd met. I hoped that our report would bring greater attention to their plight.

Inside the smoke-filled café, we talked about going to the Tumen River, which forms the border between North Korea and China in this region. Days before, we'd filmed at the bridge in the city of Tumen, one of the official border crossings. But North Korean citizens don't have the luxury of simply walking across the overpass if they want to visit China. They cannot freely leave the country, and traveling abroad is reserved for the highly elite, who must obtain special clearance from the government. Defectors must take a different path if they want to get to China, traversing the waters separating the two countries. We wanted to film at the river to document

somewhere inside north korea / 11

this well-used trafficking route, one that in the wintertime is frozen, making it easier for defectors to cross. I thought about Ji-Yong's story and how she, like so many other North Korean defectors, had braved the ice-cold waters to escape their country's poverty, only to end up being used and exploited.

Throughout the night, our guide had been getting calls on his black cell phone. He had two phones, one black and one pink. He claimed the black one was used to communicate with his contacts in North Korea. He said he'd been talking to an officer in the North Korean military and was trying to determine if any defectors were crossing over and if we might be able to interview them. He also suggested the possibility of chatting with a North Korean border guard while standing on the frozen river. He said he had taken journalists to the area before, and they had been able to make small talk with some of the lackadaisical soldiers.

We wanted to get closer to a part of the Tumen River where defectors typically cross, so late that night we drove about an hour to the city of Tumen. We checked into a hotel and planned to head to the river the next morning before sunrise. We didn't intend on staying at the river long because we wanted to get back to Yanji to catch an afternoon flight south to Shenyang, where we would continue on with our shooting schedule.

I looked out the window of my room and could see the twinkling lights from a North Korean village off in the distance. We'd been told that at different times, the whole area across the border goes pitch-black from electricity shortages. An hour later, I peered out the window again and could not spot a single light on the other side. Satellite images of the Korean peninsula at night paint a stark picture of a brightly illuminated South Korea compared with the North, which is bathed in utter darkness. It's as if a child had taken a black marker to the upper half of the peninsula.

I set my iPod to wake me up at 4:00 A.M. It was already 1:00 A.M.

by the time I got into bed. I figured I'd plow through on little sleep until we were on our flight later that afternoon, when I could take a nap. By 4:15 A.M., the time our team had arranged to meet, I was in the lobby. After about five minutes of waiting groggily, I decided to knock on everyone's doors to rouse the group. Our guide had been adamant about our filming early because he figured there would be fewer people around. I rapped on Mitch's door; he was gathering his belongings. But when I knocked on Euna's and our guide's doors, no one answered in either room. I began pounding on Euna's door and shouting out her name. Confused and worried, I went down to the lobby and had the woman at the front desk call her room. After several rings, Euna finally picked up. She explained that she and the guide had gone out to the river to try to get some evening shots. They had been out late, which is why they overslept. She called the guide's room to wake him up. We were out the door of the hotel fifteen minutes later.

On our way to the river, our guide, who lived in the area, stopped off at his home to pick up a warmer jacket. The morning chill was numbing. I had on multiple layers of clothing under a coat Lisa had loaned me, along with a thick scarf and gloves. Despite the weight, I was glad to have on my sheepskin-lined leather boots. Our guide emerged wearing a long black coat. At first I didn't notice anything odd about the jacket, but when he turned away from me I spotted the word *police* written in English on the back. A badge on the sleeve revealed what appeared to be a Chinese police patch. I felt slightly uneasy with his disguising himself as a cop, but I figured he'd done this before and knew what he was doing. I took his attire to be a precautionary measure, one that he had used on previous excursions to the river with media to better avoid detection.

As we drove to the river, our guide told Euna in Korean that he had decided to go to a different location than the one he had previously mapped out. There was a spot a little farther down the way that he thought would be better for us to film. I didn't think much of this change in plans. The guide was from the area and knew the vicinity well. Foreign journalists place a lot of trust in their local fixers or guides, and I didn't feel any reason to question his decision.

Minutes later our car pulled off the pavement onto a dirt path. Our guide drove through large patches of dried grass and weeds until coming to a stop within the brush.

The river wasn't immediately within sight when we got out of the car. We had to walk through the grass and over a small mound of dirt to reach it. The sun was just beginning to peek through a thin layer of fog as we made our way toward the border. The only noise was from our own footsteps and breath. When we arrived at the river's edge, we saw that it was frozen. That's what we were hoping for. Knowing that many defectors attempt to cross the border in the winter months so they can walk across the ice rather than navigate through the rushing waters, we too intended to set foot on the frozen river to give our audience a glimpse into this world.

Our guide made his way onto the ice and we followed. When I placed my boot onto the frozen river, the sound of crackling ice sent chills throughout my body. Though the temperature outside was bitterly cold, spring was settling over the region, and parts of the river snapped under my feet. I feared the ice was not too far from breaking. I began to tiptoe ever so carefully, feeling the crunch of icicles with each step. I held my breath, somehow convincing myself that this made me feel lighter. As Euna followed me on the ice, she began filming the area with her digital video camera. Mitch pointed his camera at me as I narrated where we were. I motioned toward North Korea on the other side of the narrow river. From here, I could see why the area has become a popular crossing point—the width of the river seemed to be the length of an Olympic-size pool.

Our guide then let me hold his black cell phone, the one he used for smuggling operations. I explained how smugglers like him call their North Korean connections and do business. Euna asked me to walk along the ice so that she could get some shots of me. I proceeded cautiously, walking parallel to the riverbank. Until this point, I never thought I would be setting foot on North Korean soil.

There wasn't a single sign or fence to indicate the international border, but we knew North Korea was on the other side of the river. Our guide began walking across the ice toward North Korea while making several low-pitched hooting sounds. His actions startled me at first, but I assumed he was trying to make contact with the border guards he knew. He continued walking and motioned for us to follow him. We did, eventually arriving at the riverbank on the North Korean side. Off in the distance was a small village, which our guide explained was where the North Koreans wait to be smuggled into China.

I was nervous. I could tell we all were. We'd never planned on crossing the border, and just as it began to sink in that we were actually in North Korean territory, we knew we needed to leave. We weren't on the edge of the riverbank for more than a minute before we turned around and headed back across the ice to China.

Midway across the river, I heard yelling coming from downstream. I looked in that direction and saw two North Korean soldiers sprinting toward us with rifles in their hands. Immediately I felt a wave of panic and started running. I no longer cared that the ice might rupture. I just wanted to get away fast. When I was just two steps from the riverbank on the Chinese side, the ice cracked below my left boot causing it to slip into the frigid water. Fearing I might sink to my death, I quickly threw my body onto Chinese soil, pulled my leg free, and continued to run.

I turned to see how far away the soldiers were and determine if they were going to chase us after we reached China. Euna and our guide were about eight yards behind me, with the soldiers closing in on them. Mitch, an avid runner, was around six yards ahead of me. I remembered that I still had on the wireless lavalier microphone and that Mitch could hear me through his headphones.

"Mitch, keep filming," I said as I continued to run.

If we were apprehended, I wanted him to have it on tape that we had been taken in Chinese territory. Mitch turned back toward me, pointed the camera in my direction, and then disappeared over a small hill.

With each step, my foot that had fallen into the cracked ice felt heavier and heavier, like a weight pulling me into the ground. "Run, Laura, keep going," I said to myself. But as in a dream when the force of the world seems to be pinning you down, I found myself falling, unable to budge.

"Euna, I can't move," I said to her as she approached.

She stopped beside me and knelt down to help. Seconds later the two soldiers were on us, with their guns pointed. To this day, I live with the guilt of wondering if Euna would have been able to outrun the guards had she not stopped for me.

Our guide, who had been able to elude the guards as they encircled us, walked back cautiously in our direction, but not close enough for the soldiers to reach him. He told Euna to take out some money, which she did, and offer it to one of the guards. Pausing for half a second, the soldier next to Euna seemed to consider taking the few hundred Chinese yuan, the equivalent of about one hundred U.S. dollars, but his comrade standing above me would not be persuaded.

"Take me instead," our guide pleaded in Korean. But when the soldier tried to reach for him, the guide dashed off. The soldier next to me grabbed my bag and noticed Euna's small video camera, which I had been trying to cover with my leg. The red record light was on.

"Please, please, please," I called out in English. "We're sorry. We're foreigners."

I knew they couldn't understand a word I was saying, but I was hoping they would sense an innocence in my tone and feel sympathetic. Furious, the guard holding Euna grabbed her camera and backpack and told us to get up and walk.

"They want us to go across the river," Euna translated.

"Euna," I said nervously, "tell them that I want to walk, but I can't because my foot is numb from falling into the water."

Though I still had feeling in my foot, I began to hit my boot so it would appear that my leg was truly immovable. I was trying to buy as much time on Chinese soil as possible. I figured that as long as we were in China and not on the North Korean side, we might have a chance.

The soldiers were intent on taking us across the river and began pulling us toward the ice. We frantically tried to cling to bushes, the ground, anything that would keep us in Chinese territory, but we were no match for the angry soldiers. The one guard standing above me was particularly ferocious. His grip was strong and his eyes piercing. To let me know he was serious, he kicked my jaw and shoulder with his heavy black boot and then delivered another crushing blow to my shoulder. I felt my neck snap from the first kick, and my whole body went numb from the second one.

Before we knew it, the soldiers were dragging us back onto the frozen river. Euna and I scrambled to grab each other. Not wanting to be separated, we grasped each other's hands. The soldiers violently ripped us apart and continued to haul us across the ice.

I tried to make the weight of my body as heavy as possible as I lay there on the frozen river. The soldier who was dragging me by one of my arms looked down at me with a fiery intensity, his eyes burning with determination.

"Please, please, we're sorry," I yelled, hoping some Chinese border guards might hear the commotion and come to our rescue. "Mitch!" I screamed, wondering if he was still within radio frequency and could hear my voice. "Help us. I think we're going to die."

I saw the soldier's boot coming for me again, this time

pounding the right side of my face. I could feel my body writhing from the pain.

"I'm so sorry, I'm so sorry," I screamed, looking up into the soldier's cold eyes. Seething with anger, he raised his rifle. I froze in terror. *This could be the end,* I thought. In a flash, he struck the butt of the gun down against my head. Immediately, I fell into a daze.

I'm not sure when I regained consciousness, but when I did, I found myself walking behind Euna on the top of a hill above the river, heading into a tunnel. My head was still in a fog. *How did I get here? Was this really happening?* The air was cold, crisp, and dead silent. Light emerged as we left the darkness of the tunnel and descended onto a small army post. I remembered the microphone that was clipped to my scarf. Fearing the soldiers might think I was transmitting messages back to the United States, I subtly pulled the wire down through my sweater and tucked the microphone into my pocket.

We were taken into a small room, where the guards handed over our belongings to a commanding officer. We were then escorted back outside and made to wait. The post was little more than a dirt clearing, where I assumed military training took place. Several curious, wide-eyed soldiers surrounded us. In any other situation, I might have attempted to make a friendly connection by offering a smile or "hello" in Korean. For the past decade, I'd worked in dozens of countries, many of which have poor relations with the United States, yet I have always been able to establish cordial, sometimes even warm, connections with the people. But this wasn't just any foreign country. So little is known about what actually goes on in North Korea. The only thing that became immediately clear to me was the deep-rooted hatred North Korea's government has for the United States. I had to remind myself that as an American, I was the enemy.

I looked down at the ground, trying to seem meek and respectful. It was as if I had entered a parallel universe. *Would I ever see or hear from my family again?* I wondered. *Could this be my last day alive?* The combination of fear and sadness engulfed me and made me tremble.

After a ten-minute wait, we were led out of the post. The same two border guards who had apprehended us held our wrists tightly as another soldier led the way. We followed a narrow trail through dry grassland. Along the path, we saw a couple of men who looked like poor farmers or peasants. They were at least a full head shorter than me and emaciated. Their skin was dark and weathered. I could tell they were curious about us, but they averted their eyes as we passed. My heart sank with each step as we headed farther and farther inland, away from China and the outside world.

◦⑆◦ **LISA**

A T TEN O'CLOCK THE NIGHT before Laura was to leave, she phoned me to see if she could borrow my light Patagonia shell jacket.

"Do you realize how cold it is where you're going?" I pressed. "There's no way that jacket is going to keep you warm enough."

Having grown up in California, we always underestimate how severe temperatures can be elsewhere and inevitably underpack or bring inappropriate attire.

"Well, I can't find my black coat, so I don't have anything else," Laura replied.

"Baby, you can't be dealing with this at the last minute," I said. "What time is your flight tomorrow?"

"The cab's picking me up at ten in the morning," she answered.

"Shit. Okay. I'll bring you my big brown parka," I said.

I woke up extra early the next morning to fight the stop-and-rarely-go 405 freeway traffic and make it from Santa Monica to Laura's house in the valley in time.

When I got there, my sister was scurrying around the house that she and Iain had been living in for less than four months. They had gotten married nearly five years ago, but they'd been together for twelve. They had been saving to buy their first home for a long time. Our parents have always been thrifty, so frugality was ingrained in us. Laura was stressed about having just ordered some custom-designed pillows that cost more than she knew she should spend. But Iain encouraged her to go for it; he wanted her to have whatever made her happy. In the midst of her frenzied packing, Laura sat down on the couch and looked at me with serious eyes.

"Li, Iain and I have just started trying to have a baby," she confided.

I was so happy for her. She went on to say that she had recently stopped taking her ulcer medication so that she could try to conceive. Iain had wanted to start a family for a while. No one was better with kids than my brother-in-law; they just flocked to him. He would allow friends' children to chase him around the pool over and over again to the point of dizziness. We'd all get tired just watching them, but Iain had endless energy. Though he loved playing with friends' kids, he wanted children of his own. But being ten years younger than him, Laura just hadn't been ready.

When I had first heard how much older Iain was than Laura, I immediately opposed their relationship.

"Are you crazy? Thirty-one?" I exclaimed. "That's way too old. You're only twenty-one."

"He looks so young, Li, you wouldn't believe it," Laura said, trying to convince me, "and plus, I really like him."

"Baby," I urged, "you have to be careful of guys like that. They just want to mess around."

I couldn't have been more wrong.

That was the fall of 1997, and Laura was a student at UCLA. While she dated here and there, she had never had a serious boyfriend before. I used to worry that she might not ever find someone, because she had never really expressed an interest in anyone. Or maybe it was just because she had never told me about it. Laura and I never kept secrets from each other, but I was always very protective of her—perhaps overly so. In hindsight, I probably wasn't ready for my little sister to start dating. As a girl, I had more crushes than I can recall. Much to my embarrassment, I was named "biggest flirt" in my middle school yearbook. I had already had a lot of experiences, and I didn't want Laura to get distracted by the boy craziness that had struck me long ago. She never would.

Iain is boyishly handsome, and many say he looks like the actor Michael Vartan with a touch of Hugh Grant. He is a rare combination of brainiac financial quant and British surfer dude. I always tease him about his recreational reading, which ranges from esoteric books about calculable formulations to ones about mathematical models. Although he could be categorized as a bona fide nerd, Iain has never lacked admirers of different ages and genders. But the most striking of his characteristics is his gentle demeanor. In more than a decade of knowing him, I have never seen Iain get angry—not once, ever.

He has a soft-spoken, kind way about him, but he's stoic and never particularly emotional—except when it comes to Laura. Their love is that of storybooks—it's the only way I can describe it. After many years, a lot of relationships grow stagnant and stale—but not my sister and Iain's. On many occasions I've caught Iain stroking her hair or rubbing her back during periods of stress. His obvious adoration of my sister and hers for him has never waned, even in the slightest.

Laura and Iain married in June 2004, seven years after they met. I had to share my best friend, but there was no one I'd rather share her with. During my maid-of-honor champagne toast at their wedding, I closed by saying, "Baby Girl, I may have been the flirt, but you got the boy."

~~ **LAURA**

AFTER ABOUT FIFTEEN MINUTES of walking along a dirt trail with the North Korean soldiers, we arrived at a second army post. It must have been no later than 7:00 A.M. It was hard to believe that the day was just beginning. This facility was slightly larger than the previous location but rudimentary all the same. While Euna was taken into a room to talk to the officer in charge, I was led through the dim sleeping quarters, which contained half a dozen metal bunk beds with thin, stained mattresses, to a small washroom. There was no sink, just a large bucket of water. On a ledge sat a couple of used, brown-stained toothbrushes. A soldier handed me a dirty rag and motioned for me to clean my face.

I hadn't thought about my injury or appearance since that moment on the ice. I touched the side of my face; my jaw was tender. It hurt to open my mouth. Dried blood from the gash on my head had caused a large chunk of hair to stick together and harden against my skin. It was difficult to peel away the hair to inspect the actual injury. I winced in pain as my fingers touched the bloody lesion for the first time. Not wanting to infect the wound with the grimy towel, I lightly wiped my face, steering clear of the injury.

I was then led into the room with Euna and the officer. There were no signs of technology, no electronic equipment, not even electricity for that matter. Euna spoke Korean to the officer in charge, telling him we were university students working on a documentary about the border region. She told him we had made an innocent mistake. I asked Euna to convey to the man that we were very sorry and ask if he could please take us to the official bridge over the river between North Korea and China so we could walk back to China. I didn't think they would, but hoped there might be a slight chance they would send us back over the bridge so the Chinese authorities could deal with us.

"Tell him we're sorry and that we could pay a fine if necessary for any inconvenience we've caused," I added.

We were made to wait outside. Euna was shivering. Her pants were soaking wet. This was the first time I noticed that she didn't have her jacket. She quietly told me she had purposely tossed her coat while we were attempting to flee on the Chinese side. She had her cell phone inside the pocket and didn't want the North Koreans to get any of the numbers that were on it. I wrapped my coat around her and tried to warm her legs by rubbing them and gently massaging them. I had a small package of trail mix in my pocket and encouraged Euna to eat some to keep up her strength. I nibbled on a few cashews and tried to remain calm.

Moments later the officer returned and in Korean explained that they would take us to the bridge. I couldn't believe what I was hearing. Were they really letting us go? He looked me over and ordered a soldier to bring me a rag to wash the blood off my parka. I took this as a good sign, thinking they didn't want the Chinese authorities to see that I'd been beaten on Chinese soil. He looked at my head and inspected my face. Fearing I hadn't cleaned myself up well enough, I had Euna tell him that there was a hat in my bag, which I could put on to look more presentable. He allowed me to retrieve the white wool cap, which I put on to his satisfaction. The officer seemed trustworthy; there was a kindness in his eyes. But I was still skeptical of his intentions.

A soldier on an old military motorcycle with a sidecar approached, and we were told to get in. We were given our belongings. Wanting to hold the officer to his word, I asked Euna to see if the man could accompany us. He said it wasn't possible, that he needed to stay at the post.

"Don't worry," he said reassuringly. "You'll be fine."

"Thank you, thank you," I replied in Korean.

Another soldier hopped on the back of the rickety motorcycle, and the driver tried to start the engine. I looked around the dusty base. There were no other vehicles in sight. This dilapidated motorcycle appeared to be the only transportation available. Three or four attempts later and the cycle finally began to roar. We were off. Euna was sitting in the front of the sidecar. I curled up behind her, bracing myself against the brisk morning air.

For the next fifteen minutes, we continued down a bumpy dirt road, rarely seeing any other vehicles. We passed a small village consisting of simple adobe buildings. There were a few people riding bicycles, but most were walking. Despite the frigid weather, the villagers were not wearing heavy overcoats. They had on simple dark, drab garments, which matched their gloomy expressions.

I wasn't sure of our location, but it seemed we were headed in the opposite direction of the Tumen Bridge that links North Korea and China. Still, my instincts told me we were traveling parallel to the river, which gave me some relief. I figured they must be taking us to a different, closer border crossing. A military truck approached us, and someone inside motioned for our driver to stop. Their conversation was inaudible, but thankfully, we were soon on the road again.

Suddenly we made a left turn, heading away from the river. This is when I knew immediately we were not going back to China. I grabbed Euna's shoulders, rubbing them as if to warm her, but hoping she would take this as a signal that something was very wrong. We ascended a path and pulled into a larger military base.

We were ushered into an empty room where three officials were waiting. We all sat on the linoleum floor, which was slightly heated by an underground wood- or coal-fired furnace. This was the first time we'd experienced any kind of warmth, and I pressed my hands to the floor to restore the feeling in my fingers.

The officers proceeded to look through all our belongings, showing particular interest in our equipment and money. I handed over the microphone I'd been keeping in my pocket. We had roughly three thousand dollars in our possession, consisting of South Korean, Chinese, and U.S. currency. At each place where we were held, the officers had counted our money and noted how much of each type of currency we had. Here, they meticulously counted the cash again. I suspected they wanted to make sure no money was missing or had changed hands from location to location. They leafed through our passports, pausing to look at each of the dozens of visas in my booklet.

"Why do you have so many visas?" asked one of the officers.

"My family really likes to travel," I replied nervously with Euna translating. He didn't seem convinced.

Another officer picked up the receiver from a telephone that was on the floor in a corner and tried to place a call. This was the first bit of technology I had seen in the three different locations in which we'd been held. To his frustration, there was no connection. He tapped on the receiver button repeatedly but was unsuccessful. I wondered if he was trying to contact higher authorities or officials in the capital, Pyongyang. The out-of-date-looking telephone and lack of connection seemed to be signs that we probably didn't need to worry about the room being bugged or electronically monitored. I wanted to be with Euna alone so we could speak more freely and figure out a plan.

So far, Euna had informed the officers that we were students working on a documentary project about the border and trade between China and North Korea. We knew the issue we were really covering, North Korean defectors escaping from their country's poverty and brutal government, was particularly sensitive and that the missionary groups that had been aiding us were not liked by the North Korean regime. I began to think about what evidence we had that might compromise our sources and interview subjects, or reveal what our true purpose was in the region.

An officer pulled out the digital still camera that was in my bag. He handed it to me and asked me to show him the photos. I remembered the pictures of North Korean women defectors I had taken. One was of a girl who had fled from North Korea and was lured into the online sex industry in China before being smuggled into South Korea by missionaries. The other was of a woman who had been forced to marry a poor farmer in China. While that photo only showed the back of the woman's head, I didn't want to take any chances. I nervously deleted these pictures before showing the officer some of the benign ones, such as me enjoying a traditional Korean meal in Seoul.

We were then taken to another building. But before we left the first facility, we were blindfolded with two bandanas I had in my bag. I'd become accustomed to carrying bandanas on my trips because of their versatility—they can be used as handkerchiefs, hair wraps, or protective cloths. Now my own bandanas were being used to keep me prisoner.

Two female soldiers led us across a courtyard. As we stumbled from one building to the next, I could hear military drills being conducted nearby. The sounds of boots marching to a beat and the cadence of the soldiers' voices sent my heart thumping with trepidation.

Our blindfolds were removed after we entered a room much like the previous one. We were told that someone was coming to take us to another base. Now that it was clear we were leaving these officers' jurisdiction, the air seemed to become a little more relaxed, and for a brief period we were left alone with our belongings. With soldiers right outside our door, we scrambled nervously to destroy whatever evidence we thought might get our sources, interview subjects, and us in trouble.

I told Euna I had deleted some pictures from my camera.

"What should I do with my videotapes?" Euna asked.

"I don't know," I replied, trying to recall what was on the tapes.

Two of them contained an interview I had conducted with a recent North Korean defector. He, unlike the women we'd spoken with, had fled because he was upset with North Korea's political system. While Euna's tapes did not reveal the man's identity because she had only filmed the lower part of his body, the types of questions I had asked him could be quite damaging to our situation. Euna proceeded to rip the ribbons on the tapes so they would not be viewable.

I had a small notebook, and several of the pages inside contained interview questions for Pastor Chun Ki-Won and a professor in Seoul, two men whose work is considered subversive by the North Korean government. I carefully ripped the pages out of the notebook. Euna told me to give her a page. She crumpled up the paper and put it in her mouth, chewing and swallowing. I followed her lead. Fearing I might exacerbate my recurring ulcer, I ripped the other page up into small pieces and put it in my pocket. Later on, I asked a guard if I could use the toilet, which was an outhouse on a raised platform. I wrapped the small bits of notes in a sheet of toilet paper and dropped it into the trough below.

In the time remaining, Euna and I discussed how we would continue on with our tale about being students. We decided we would tell the authorities we were graduate students at the University of California, Los Angeles, film department, and that Mitch Koss was our professor. So far, we hadn't met a single person who spoke English. We were far from the capital, and I hoped it would be difficult to get a translator and that they would allow us to remain together. That way, we could keep our stories straight.

The day was far from over, but already it seemed like the longest one of my life. My head was pulsating. I was so fatigued that my worries and nervousness subsided. All I wanted was sleep. I closed my eyes for a brief moment, before forcing myself awake. I became concerned that because of my injury, if I dozed off, I might fall into a state of unconsciousness. I pinched myself to stay alert. I tried to comfort Euna by telling her we would be okay, that North Korea had more to gain by keeping us alive than dead. I told her I didn't think we'd be sent to jail, but would probably be placed under some sort of house arrest. A few hours later these words would come back to haunt me.

So far, the people we'd encountered had seemed suspicious of us but relatively compassionate. I feared being moved to another location where the people might not be as kind. As dusk approached, new authorities arrived to transport us to another facility. By this time, we were supposed to have been on an airplane heading to another Chinese city. Instead we were prisoners inside North Korea. We were blindfolded again, handcuffed, and crammed in the backseat of an SUV between two officials. We were told to look down and not speak. Silence ensued.

We traveled over bumpy terrain for what seemed like thirty minutes before arriving at the place where we would end up being interrogated and held for the next three nights. Euna was taken out of the car first. We'd been together all day, able to console and confide in each other. Now we were separated, and a sense of anxiety rushed over me. A soldier removed my handcuffs, pushed my head down, and led me into the building.

We'd been transported to a jail. Before entering the building, the soldier motioned for me to take off my shoes. He then unlocked a door that led into a small, dim area that housed a row of four cramped cells. The soldier removed a heavy lock from one of the cells, opened the door, and directed me into the dismal five-by-six-foot chamber. The deep echo of the door shutting and the lock clashing up against it made my skin crawl. Rather than having metal bars that allow one to see into each cell, these chambers were fashioned with heavy metal doors. There were two postcard-size slots in each door, one at the top for a guard to look through, and one at the bottom through which a small bowl of food could be placed. If the slots were closed, the room was pitch-black. Fortunately, a sliver of light entered my cell through an opening in the upper slot, and I could make out a thin pallet of wood on the concrete floor along with a pillow and two blankets. I sat down, buried my face in my hands, and began to sob.

I thought of Lisa, my parents, and my husband, Iain, and the horror they must be feeling not knowing where I was or if I was even alive. While in Seoul and China, I had managed to speak with Iain via webcam. But the time of our usual chat sessions had long passed. By now he must know that something was wrong.

CHAPTER TWO

# scrambling for answers

⊰⊹⊱ LISA

**P**AUL'S PHONE JOLTED US AWAKE. I had turned my cell-
phone ringer off before going to bed because I had been
woken up by early calls from the East Coast the previous two
days in a row. The clock read 2:30. It was the morning of March 17.
Paul's initial grogginess quickly turned to seriousness as he passed
the phone to me. It was Laura's husband, Iain.

"Laura's been abducted by North Korean border guards," he
said.

I couldn't respond. I just froze.

Iain explained that the producer/cameraman Mitchell Koss
had evaded capture and was able to get a call out to his wife, who
contacted one of Laura's work colleagues, who then called him.

"Have you called my mom?" I asked.

"No," Iain replied. "Should we tell your parents yet?"

"Yes, we need Mom to start reaching out to Chinese government offices," I said, meaning that we needed my mother's proficiency in Mandarin Chinese. "I'll call the U.S. Embassy in Beijing right now."

I also left an urgent message for my friend Richard Holbrooke, the U.S. special representative to Afghanistan and Pakistan and the most senior diplomat I knew. Richard had been a foreign policy mentor of mine since my former boss at ABC TV's *The View*, Barbara Walters, introduced me to him in 2001. Richard helped broker the peace agreement between warring factions in Bosnia in 1995. He was also U.S. ambassador to the United Nations in the Clinton administration and had worked extensively in Asia throughout his career as a diplomat. I knew he was very close to Secretary of State Hillary Clinton, so I called him, hoping he could at least tell me what to do next.

By 4:00 A.M. I was on the road, headed to my mom's house in the valley, which is about a twenty-minute drive from my home in Santa Monica. Iain, Paul, and I would practically live at her house for the next few months. As soon as Dad got the news, he flew down from his home in Sacramento to be with us. We based ourselves at Mom's house because hers was the biggest and had the most bedrooms. We brought our essential items, set up our computers, and turned the place into our own Ground Zero in our efforts to get Laura home. We reached out to Euna's husband, Michael Saldate, and their four-year-old daughter, Hana, and told them to make our home theirs. Euna's parents were in South Korea and her sisters lived in other U.S. cities, so Michael and Hana became extensions of our family. They would spend many weekends with us at my mom's house. Though we took the lead on release efforts, we wanted Michael and Hana to know that we would be working tirelessly on both Laura's and Euna's behalf.

We also phoned Joel Hyatt, the CEO of Laura's employer,

Current TV, and asked if he could wake up Current's chairman, former Vice President Al Gore, right away. If this became the international incident we thought it would be, we needed Vice President Gore to flex his political muscle. As soon as he got the news, Gore got my phone number from Hyatt and called me; it was about 6:00 A.M. His deep southern voice was comforting. My sister had a very good relationship with the former vice president and I know she revered him, but except for a couple of brief exchanges at special events, I didn't really know him. He put me at ease right away. With calming authority in his voice, he told me our family should trust that he would take control of the situation. This was a huge relief for us.

"I've been briefed by Joel about what's going on," he said. "As soon as people start getting into the office in Washington, I will start making calls right way. You call me anytime, and tell your parents that we are going to do everything we can to get Laura and Euna back as quickly as possible."

Our conversation was not a long one, but I felt lucky to have the former U.S. vice president on our team. If anyone could open doors, surely it was Gore. He gave me all of his contact information and told me to share it with Iain and everyone else in the family. But there was something of vital importance that he suggested we not do: talk. He strongly advised us not to say a word to anyone, especially the press. At this point, no one knew what had happened to Laura and Euna. That wasn't surprising, because so little is known about what goes on inside North Korea. He warned that we did not want to do or say anything that might inflame those holding my sister and Euna. We had to exercise extreme caution.

"The next forty-eight hours are crucial," he urged. "We're not dealing with a normal government, and we have to be very, very careful."

LAURA AND I HAD faced unpredictable situations before throughout our careers. We both started working as journalists when we
were very young, and we'd traveled to dozens of countries, some
of which were unstable during our visits. When I was eighteen
years old, I was hired to be a correspondent for a news program
that was seen in middle and high schools across the country; it
was called Channel One News. Channel One routinely sent correspondents all over the world to cover stories that network news
wasn't covering: we reported on the civil war in Algeria, globaliza-
tion in India, sex slavery in Saipan. For a kid who'd never gotten a
chance to travel, Channel One opened the world to me. It exposed
me to distant lands and foreign cultures and broadened my sense
of humanity in ways I would have never been able to experience
otherwise.

Early on in my tenure at Channel One, I started working with
the man who became my mentor and eventually my sister's, Mitchell Koss. This was the early nineties, and our style was drastically
different from what traditional news was doing. First off, Mitch
became the cameraman as well as serving as producer. So unlike
most news crews, which required a sizable number of people, we
were compact and easily mobile, and because it was just the two
of us, we were cheap. Our style was casual and experiential, and
we immersed ourselves in the stories we covered, giving viewers a
very accessible kind of reporting. We went to the front lines of the
war in Afghanistan, to the cocaine-producing jungles of Bolivia,
and to Tibet, where twice we went in posing as tourists.

We were able to do this because we didn't have to shoulder
the weight of a big, well-recognized news organization. Journalists
on official visits to China must provide a detailed list of intended

shoot locations and interview requests, and they have to provide a letter of invitation from a Chinese organization. If granted, an escort from the government is assigned to monitor every step and word inside the country. We knew that if we acquired official journalist visas from the Chinese government, we would not be able to cover the kind of story we wanted to cover in Tibet. As tourists, we could capture a much more realistic, nonpropagandistic look at the reality of life there. When we stopped in a Buddhist temple, two young monks approached us and whispered that a year ago Chinese police officers had taken a few monks away from the monastery, and they'd never come back. I was struck and humbled by how brave these monks were to talk to us. They were literally risking their lives to alert us to what Chinese officials had done to their brethren. We would never have had such candid conversations had we gone in as journalists. Going in officially would have inevitably skewed the story in one direction, China's.

During my seven years as a reporter at Channel One, Mitch and I crisscrossed the world several times over. I relished every second of my job at Channel One until it was time to get a TV job that people outside high school would see.

That's when, in August 1999, I was offered a job as cohost on the daytime talk show *The View*. Later that year, Laura became a researcher at Channel One. After working with me for so many years, Mitch had become a close family friend. So it wasn't surprising that when she started at Channel One, Laura began to work with Mitch as well.

∾〜 **LAURA**

WAS A COLLEGE STUDENT at UCLA when Lisa's career began to take off. I looked at her work with both pride and envy. She made me want to know more about what was happening in the world; she made me care about things I would never have known about. Her reports opened my eyes to what was happening in Cambodia, Iraq, and Kazakhstan, places that few people were paying attention to. I was left not only wanting to know more, but yearning to travel the world myself and investigate new situations.

Working at Current TV, a new cable network that targeted the young adult demographic, gave me, along with a team of journalists, the opportunity to raise awareness about critical but underreported issues. Our department was called Vanguard because, at a time when most news networks were cutting back on the number and kinds of stories they were covering internationally, we sought to fill the void.

Like my sister, I had become accustomed to reporting in hot spots. Just a few months earlier I had taken on one of the more dangerous assignments of my career. Along with Mitch Koss and an associate producer, I'd traveled to different cities in Mexico to report on the escalating drug violence that was paralyzing the country. Warring cartels were battling with each other and the Mexican government. In Ciudad Juárez, which borders El Paso, Texas, we rode with a local journalist to places where half a dozen gruesome homicides had taken place. A cemetery in the city of Culiacán revealed an endless patchwork of graves representing a generation of young men murdered in the narco-war.

Lisa and I worried about each other during every risky assignment and made sure to check in regularly. Lisa was relieved when I finally finished the Mexico project. Before I left for Asia, she told me to hurry up and get home so I could slow down and start trying to have a family. She knew I had been thinking about having a baby for a while, but the pace of my job kept interfering.

"Don't worry," I said. "There won't be any bullets flying on this story."

Now I was sitting in a North Korean jail—I hadn't expected this project to be physically dangerous at all. My biggest worry had been about protecting our interview subjects. But I never could have predicted this situation. I thought back to the conversation I had with Lisa about slowing down. The morning of my flight, I told her that Iain and I were finally trying to start a family.

"Who knows?" I said, smiling. "I could be pregnant right now." Those words rang through my head. *What if I were pregnant?*

My thoughts were interrupted by the clanging of a small tin that was being pushed through the lower slot in the door. It was a metal bowl of rice with a few pieces of kimchi, a traditional Korean dish made of cabbage and spices. My jaw was still aching from the soldier's kick with his boot; it was a challenge to open my mouth. I forced myself to take a few bites to keep up my strength. Soon after dinner, I heard a guard open the cell door that was to the left of mine. I could make out Euna's soft voice inside. A feeling of relief rushed over me when I discovered that at least she was nearby.

I had known Euna for more than four years. She'd edited a number of projects for Current TV's journalism department, many of which I had worked on personally. Together we'd sit in one of Current's cramped editing rooms and scan through the videotape from a particular shoot, trying to find just the right image that would make a scene come to life. Euna always had a photo of her adorable daughter, Hana, pinned up somewhere near her computer screen. Although this project along the Chinese–North Korean border was the most extensive one we'd engaged in together, being out in the field with Euna felt natural. While I didn't know her well personally before this trip, the events of the last twenty-four hours had bonded us for eternity.

I peered through the slot in my door and saw Euna exit the room followed by a uniformed soldier. Several minutes later, I too was taken out of my cell into a separate interrogation room. I was met by two male officials and a female translator. The first man was tall, robust, and handsome. The other was of average height and heavyset. Unlike the other authorities we'd dealt with earlier in the day, these men were not wearing military uniforms. Both were dressed in the typical North Korean attire—dark Mao-inspired suits with pins of the Korean flag—and they seemed even more intimidating. The translator had perfectly styled hair. She spoke in very broken English—between recurring coughing fits—and said she was an English teacher at the local school. I struggled to make out her words and wondered how well she understood me.

I took a seat, cross-legged, on the floor in the center of the room. The area consisted of two chairs, a glass coffee table, and a Chinese-brand television. But rather than sitting in the chairs, the officers plunked themselves down on the floor, leaning their backs against the wall. They began asking me questions through the translator about what I was doing at the river. I stuck to the story Euna and I had rehearsed and explained that I was a university graduate student working on a documentary about trade along the border.

"What other things were you filming besides being on the river?" they asked.

I told them about some of the merchants we had spoken with on the Chinese side of the border who were selling cigarettes, money, stamps, and herbs from North Korea.

After they asked some basic questions such as "How many students are in your class?" and "How many women students are there?" I knew this fabrication could not go on for much longer.

Fearing I might cause trouble for Euna if I came clean and told them what our true professions were, I decided to continue on with the lie until I had a chance to compare stories with her. I based my answers on the number of people who worked in our journalism department at Current TV.

"We have sixteen students," I replied. "Nine of them are women."

I hoped Euna was using the same logic. I later learned that she said we were the only students in our class; that it was a kind of independent study program.

"You're lying!" shouted the translator on behalf of the taller officer. "Do you know that as an American, you are an enemy of my country? Why would you want to come to my country if you were not invited? Euna Lee has been frank with us. If you are not frank with us . . ."

He then moved his hand across his neck in a slicing motion. I started to tremble and broke down in tears.

"You are being cunning!" the woman translated in a harsh, stern tone. "Don't try to gain sympathy with your weeping. Do you think we are fools?"

I looked up into the official's dark, narrow eyes. I felt the stabbing of his sharp gaze and bitter scowl. He handed me a pen and a few sheets of paper and told me to write down information about my family. They wanted the names, ages, and work history of all immediate family members and spouses.

I tried to think of a way to describe Lisa's profession. I knew they could easily find out that Lisa was a journalist and that a simple Internet search could reveal her work on a controversial documentary she surreptitiously filmed in North Korea for National Geographic Television. That information could be hugely detrimental to my situation. But I wasn't convinced that in this remote part of the country they even had the technological capability to access the Internet. "Sister—Lisa Ling Song," I jotted down, adding Lisa's husband's surname to her designation. "Profession—housewife and volunteer." I figured I wasn't lying completely. Lisa was a wife and a volunteer.

This questioning session went on for several hours. During this time, I often explained that I couldn't understand what the translator was asking or saying. I tried to play ignorant so that Euna's stories and mine would not conflict. At one point the electricity went out and the room went dark. A single flashlight provided light for the remainder of the evening. I would see that these blackouts happen multiple times in a day.

At one point I looked down at my jacket and saw fresh droplets of blood. I had forgotten about my injury. The events of that morning seemed like a lifetime ago. The wound had opened up again and my head began to throb. I remembered I had a few tablets of Tylenol in the bag I had been carrying when we were apprehended. I asked the officers if I could take them to alleviate the pain. They summoned a soldier to retrieve my bag and watched carefully as I rummaged through the contents until I found the package of pills. The interrogation lasted about six hours. A clock on the wall read 3:00 a.m. when I was finally escorted out of the room.

Back in my cell, I curled up on the wooden platform and pulled my bloodied parka over me as an extra layer of warmth. As I clutched the puffy winter coat, it was a reminder of my sister, who on the morning of my trip had driven all the way across town to see me off. She was concerned about the biting cold of northern China in March, so she rushed over to loan me her warmest clothes.

Suddenly a beam of light shone into the cell as the soldier standing watch aimed his flashlight through the slot in the door. His cold gaze sent daggers through my body. He slid the metal opening shut, and the darkness engulfed me. I was terrified to the point of breathlessness.

Then I heard two distinct knocks on the cell door to the right of me. Euna's cell was to my left; there were two others to my right. The knocking was followed by a brief silence, and then I heard two more taps, this time coming from within the cell. This went on for a few more moments. I listened closely and heard the guard move on to the next cell down and begin the knocking again. Until this moment, I hadn't known that other prisoners were being held with Euna and me in the four-unit jail. This eerie knocking ritual between guard and prisoner seemed to be a system the guards used to make sure the inmates were still alive. It was either that or some strange way of inducing sleep deprivation because it continued throughout the night.

I wanted desperately to communicate with Euna. I decided to test how strict the guards were and to see if I could converse with her from cell to cell. I tapped the metal door and a guard opened the slot.

"Thank you," I said respectfully in Korean. Then, a bit louder, I said, "Euna," hoping she could hear me.

"Yes? Laura?" she replied.

"Euna classmate, would you please tell the guard I have a bad stomachache, Euna classmate, and could I please use the toilet, Euna classmate?"

I was trying to let her know I had stuck to our original story about being students, and I wanted to see if she too had stood by the plan. But other than hearing her translate my request to the guard, who unlocked the door to let me use the bathroom, I couldn't tell if she had told them we were journalists. Euna and I needed to tell the Korean authorities the truth soon. If we continued to lie, we might be viewed as spies, which was the worst possible scenario.

I started to wonder what had happened to Mitch and whether he had made it back to our hotel safely. The terrifying events of that morning flashed through my mind, and I kept replaying the image of Mitch vanishing over the mound. I didn't harbor any anger toward him. Everything on the ice happened so quickly; we were all completely petrified. I thought of Mitch's wife and two kids and felt relieved that he was not in our situation. I even saw it as a blessing that he got away, because he would be able to contact our families. I desperately hoped he was able to alert them about what had happened.

OMETIME IN MIDMORNING, I finally got a call from Mitch. He said that immediately after Laura and Euna's seizure, he had turned himself in to the Chinese authorities with the hope that they would do something to help get the girls back. They interrogated him for about fifteen hours, and he had just been released from custody and allowed to go back to his hotel in Yanji. It was close to midnight where Mitch was, but he and I spent more than an hour on the phone going through the details of exactly what had happened on the river in those early hours.

Mitch said that Laura, Euna, and he had followed a guide who brought them to the Tumen River. When journalists work overseas in unfamiliar places, we often hire fixers whom we trust to take us where we need to go. In my sister's case, they did exactly that. They hired a man who had worked with news crews before to take them to see the border between China and North Korea. Mitch told me that the guide led the team across the border into North Korea for no more than a minute before heading straight back to the Chinese side of the river. He and the guide were able to outrun the North Korean border guards once they reached Chinese soil. The guide had also turned himself in to Chinese authorities, but Mitch hadn't seen him since.

"Did you definitely cross the border, Mitch?" I probed, even though he had already explained that they had.

"It was so fast, but yes, we crossed briefly," Mitch said.

"Are you sure the girls had reached China when they were seized?" I asked.

"Yes," he said definitely.

This was a big deal. Mitch confirmed that the team did cross the border into North Korea for a few moments, but he was also certain the North Korean soldiers entered China's sovereign territory in order to capture my sister and Euna. This gave me hope that perhaps we could get the Chinese government to intervene and help us.

Then Mitch said something that struck me.

"What do you mean, 'the guide started to make hooting noises'?" I probed.

"I don't know why he was doing it," Mitch responded. "It seemed like he knew what he was doing."

I couldn't stop thinking about this. Why would their guide make any loud noises at all under such delicate circumstances? I wondered whether the team had been led into a trap. I continued to press Mitch about the possibility that the guide might have knowingly led them across the border into North Korea.

"I just can't imagine that could have happened," replied Mitch.

"Do you think she's okay?" I asked.

"I don't know."

We kept going over what had happened again and again, even though it was after 1:00 A.M. in China where Mitch was. I just couldn't let him off the phone. I kept asking the same questions. I needed every detail. I thought that if I asked in a slightly different way, he might remember something he hadn't told me. I could tell he was exhausted; he had been up since 4:00 A.M. the day before. At a certain point, I realized that he had nothing more to tell me. After he ran, he lost all contact with Laura and Euna. My sister was still wearing the wireless microphone when she was captured. The last thing Mitch heard her say through his headphones was "I'm sorry. I'm sorry."

I had known Mitch for nearly twenty years, and we spent more than five years working together exclusively at Channel One. I had shared some of the most deeply personal moments of my life with him. He was my mentor and friend, and I will always regard him warmly. But I couldn't stop wondering if things might have been different if he hadn't left them. I couldn't focus too much on this. What happened happened. Now we all collectively needed to figure out how we were going to get the girls out. If Mitch hadn't run, it might have taken days to learn of their capture.

Later that morning, Richard Holbrooke returned my phone call. He had so much going on with his Afghanistan assignment that I was nervous about bothering him. He told me right away that he had just been with Secretary of State Hillary Clinton and that she was calling an emergency meeting to talk about our situation later on that day. He then said something that at the time was very reassuring to me. "Look, Lisa," he exclaimed, "I've seen many examples of government inaction over the years, and this is not one of them."

News of Laura and Euna's detainment broke a full two days after they were arrested; we had managed to keep it quiet until then. A South Korean newspaper was first to report the news, and it was immediately picked up by press the world over. News sites were rife with variations of the same headline: "American Journalists Reportedly Detained by North Korean Border Officials on the Chinese–North Korean Border."

The morning after the news came out, I got a call from my boss, Oprah Winfrey. I work on her show as a field correspondent. She asked if she could do anything to help. Reality suddenly struck when it dawned on me that even the most powerful woman in the media couldn't do much. For that matter, there was little my country—the most powerful in the world—could do either. We were at the mercy of a government that answered to no one but itself.

My sister was being held inside North Korea at the worst possible time. For weeks, North Korea had been saying it intended to launch what it called a "peaceful" satellite. Much of the rest of the world, however, charged that the North Koreans were trying to reignite their country's ballistic weapons program. Japan, for instance, had been threatening to shoot down any satellite or weapon that entered its territory. North Korea said it would consider any action by Japan an act of war. Tensions were growing by the day.

On top of it all, America had a brand-new president and secretary of state, whom pundits and conservative talk show hosts were ardently watching for any signs of weakness. I wondered if having two young American women in captivity would complicate things for the U.S. government. I knew the United States had a policy of not negotiating with terrorists for hostages, but what if a *government* was holding American citizens? The problem was that this was no ordinary government. North Korea is considered one of the most unpredictable regimes on earth.

Within the first twenty-four hours of Laura and Euna's absence, both of our families were introduced by phone to Kurt Tong, director for Korean affairs at the State Department. He would be our main point of contact for whatever diplomatic efforts ensued between the United States and North Korea. But our family had no idea how complicated that would be.

~∽ **LAURA**

IN THE MORNING, BREAKFAST was brought to our cells. Rather than the previous night's meager fare, this was a more substantial meal consisting of rice, half a hard-boiled egg, tofu, and kimchi. The guard also let me eat with my cell door slightly cracked open to let in more light. I guessed that the prisoners to my right were not receiving the same type of treatment. The only times I heard the guards open their cell doors was once in the morning and once at night so that they could use the toilet. I wondered if this was a good sign, an indication that they might let us go. Or were we being treated differently because it might be our last meal?

I was relatively confident that the guard standing watch could not understand English, so I decided to try communicating with Euna again.

I raised my voice, hoping she would hear me from her cell. "Euna, did you tell the officials that we work for Current?"

"No," she replied.

"I think we need to tell them the truth," I said. "We need to tell them we work for Current. Otherwise they are going to think we are spies."

"Okay," she replied softly.

To avoid arousing any suspicion, I asked Euna to tell the guard that my head was hurting and to ask if I could please see a doctor.

"And, Euna," I said, "when the officials come, please tell them I would like to be with you when we confess. The other translator does not understand what I am saying, and I am afraid she is misinterpreting what I mean."

A doctor arrived shortly after this exchange, and I was taken to a cold, dank room. I asked if Euna could join me, which they allowed. The doctor was an expressionless man. I could smell the strong cigarette odor on his breath and clothes. He proceeded to examine the wound in a very matter-of-fact way. He didn't seem bothered that his inspection of the cuts was going to cause me pain.

I squeezed Euna's hand tightly as he poked at the sensitive areas. My hair was covering the gash, so he had to cut off a patch to better see and clean the lesion. After tossing the blood-encrusted locks onto the concrete floor, he opened the wound to see if there was any infection. My shoulders and brows contracted as he probed the area with a metal tool and used a cotton pad doused in alcohol to sanitize the gash. After wrapping my head with gauze, the doctor asked me if I had any other injuries. Euna translated, and through her I told him

I did not have any sensation on the right side of my nose and that my legs were very sore. I rolled my dirt-stained jeans up and saw my legs in the light for the first time. Deep purple welts covered the entire length of both legs. No wonder it had been so painful to walk and bend down. The doctor gave me some ointment to put on the affected areas. He also gave me a raw egg that was still in its shell and told me to rub it over my black eye and nose area to reduce the swelling. The egg was cold simply from being out in the frigid air, and the cool sensation of the shell pressed up against my skin provided some momentary relief. I was grateful for the doctor's attention and for the chance to be out of the claustrophobic cell. But soon enough, Euna and I were separated and sequestered once again.

Not long after that, I heard the lock on Euna's door open. She was being taken to another interrogation session. I waited nervously, knowing she was about to tell them the truth about our jobs. Then my turn came. I was brought into a room where I saw the same two officials who had questioned me the night before. I was relieved to see Euna sitting on the floor.

"I've told them that we work for Current, and that we were producing a report about the border region," Euna explained. "They want to ask you some questions."

I was glad Euna was translating. This way I could be sure our responses were aligned and that my words would not be misconstrued.

The officials asked me about my job as a reporter for Current TV. I refrained from letting them know I was the head of the journalism department, to minimize my culpability. I explained that as a reporter, I had covered many different types of stories all over the world. I told them that Mitch was the producer on this story and that he was essentially directing Euna and me throughout the project.

"Have you ever had any contact with the CIA or anyone from the CIA?" one of them asked.

I remembered a segment I had worked on almost a decade earlier inside the CIA's headquarters, and the public information officer, Chase Brandon, with whom I had become friends. Were they asking this because they knew about my past work or were they trying to make sure we were not connected with the U.S. government?

"No," I responded. "I've never done anything with the CIA. We're journalists. Not spies."

I exhaled in relief when they didn't press me on that again. Compared with last night's grilling, this question session was much more restrained. It helped that we could answer many of their questions with relative ease. Thankfully, they seemed more interested in our backgrounds than in the actual story we were reporting, because we still hadn't told them any specifics about the interviews we had conducted with the North Koreans who had escaped the dire conditions in their homeland. Finally, they asked both of us to write a confession statement admitting we were journalists and that we had crossed into North Korea illegally.

Back in the darkness of my cell, I wondered how long we would be kept there. I shuddered to think that we could be sent to one of North Korea's notorious gulags. The U.S. Committee for Human Rights in North Korea estimates that approximately two hundred thousand people are imprisoned in North Korea's brutal labor camps, and that four hundred thousand have perished due to torture, starvation, disease, and execution. I considered myself lucky that I was in this dingy, claustrophobic jail and not in a hard labor facility.

The next morning one of the officials told us that the U.S. government had been informed that we were being held in North Korea. We also learned that the authorities in Pyongyang were arranging for us to be transferred to the capital. My emotions were mixed when hearing this news. I was glad to hear that our government was aware of our situation, in part because it meant that our families most likely knew as well. The idea that we had simply disappeared, leaving them no way of knowing if we were even alive, was gut-wrenching. But the thought of being moved to Pyongyang made me uneasy. It seemed that any chance of getting released back to Chinese authorities was now gone.

To begin the process of turning us over to officials in the capital, we were brought into a small office that had a computer and telephone. This was the first time I had seen working technology. We were asked to sign and fingerprint our confession statements. We pressed our right thumbs into the red ink and rolled them onto the bottom of the documents.

"Euna, please tell them I was the reporter on this project, and I bear the responsibility for our actions. If anyone is punished it should be me, not you."

"No," Euna cried. "I will not translate that. It's okay, Laura."

"Please, Euna," I continued, tears streaming down my cheeks.

"No," she said firmly.

"What is going on?" asked the officer.

"Nothing," replied Euna in Korean. "Everything is fine." We embraced each other, not knowing what was going to happen to us.

As we were being escorted back to our cells, I suggested that Euna ask if we could stay together in one cell. Surprisingly, they granted this request. I had sensed a palpable softening of these officials' attitude. They seemed to feel that since we would soon be in the hands of authorities in Pyongyang, their work was pretty much over.

Being together in one cell was an amazing gift. We were still careful to watch what we said, especially because Euna thought one of the guards could understand a little bit of English. We tried to comfort each other as best we could. We believed there was a very good chance we'd be separated again once we got to Pyongyang, so I asked Euna to teach me some basic Korean words and phrases, such as "Good morning," "Good evening," "My head hurts," "My stomach hurts," "I don't speak Korean," "Sorry," and "toilet."

I had been practicing yoga for several years, and even in this cramped space, I was able to show her a few yoga stretches. We breathed in deeply, stretched, and exhaled. With each breath, I prayed that something was being done back home to help us get out of this nightmare. Euna and I took turns giving each other massages to help ease some of the discomfort we had been feeling in our muscles and bones. We talked about how much we missed our husbands. I tried to hold back tears when Euna spoke about how much she longed for her little daughter. Though we had known each other for years, this was the most personal conversation we'd ever had together. We were both trying to stay positive, but it was hard to keep our minds from wandering and thinking that the worst might happen to us.

I noticed some Korean characters that were lightly etched into the wall. It was a kind of Korean prison graffiti.

"What does that say?" I whispered.

"I miss my mom," she translated.

I tried to imagine who had been in this cell before us, as well as who was being held in the nearby cells. Unlike us, they were not allowed to use the toilet whenever they asked, nor were they given special meals and allowed to talk to each other. But like all people who are isolated from society, we all missed our mothers.

Euna and I talked about what we should say when we got to Pyongyang. In the capital city there would be access to technology and they would know more about us. We knew we'd have to be as honest as possible. We were determined not to compromise any of our sources or interview subjects.

"Don't worry," I said to Euna. "I'm sure Al is doing something to help."

I was referring to former Vice President Al Gore, the cofounder and chairman of our company. It was his vision that had made me want to work at Current TV nearly five years ago. He and his business partner, Joel Hyatt, formed Current as an independent network that gave its young adult viewers an actual say in political and global issues. Vice President Gore encouraged our journalism department to seek out stories that were too important to be ignored.

I knew that being held prisoner in North Korea was one of the absolute worst situations an American journalist could be in, but I also knew that if anyone could get attention for us in Washington, it was the former vice president.

"I wonder if your sister might be able to do anything," Euna added. "She must have a lot of contacts."

"Yeah, but the problem is that she was in North Korea a couple of years ago. She came undercover with a medical team and did a really critical documentary about this place," I explained. "I'm nervous they're going to find out about that piece. That would not be good for me."

Euna and I spent that night in her cell. The guards gave me a blanket, which I placed on the concrete floor next to the wooden platform where Euna had been sleeping. The two of us lying there side by side took up the entire area of the cell.

The next day we were taken to a small washroom where we were told to clean up and make ourselves presentable for the authorities in Pyongyang. One of the officials gave us each a towel. "Comb your hair," he said as he handed us pink and green plastic combs with cartoon characters printed on the sides. "You can keep these as souvenirs," he said with a self-important smile, proud of his cruel joke. Shivering, we splashed cold water from a large basin over ourselves. It was the first time we'd rinsed off since our detainment. My body was so sore from the blows dealt by the soldier on the frozen river that it was still hard to move. When the icy water hit my skin, it sent a shock through my system.

In an adjoining room, a soldier brought in a low folding table.

The same two officials who had interrogated us entered the room and invited us to be seated for lunch. We sat cross-legged on the floor as the soldier served a number of dishes including radishes, tofu, eggs, kimchi, and North Korean traditional cold noodles. They filled our glasses with warm beer.

"What is going to happen to us?" I asked.

"Don't worry," the handsome official responded. "Our chairman, Kim Jong Il, is a compassionate man. Just be frank and honest and he will forgive you."

The megalomaniacal Kim is reputed to be one of the most dictatorial heads of state in the world. *Compassion* was one word I never heard associated with his reputation.

The official then went on to say that we'd probably be home in ten days. *Ten days,* I thought. That sounded like forever. After our first day, when the officer assured us we were being taken back to the bridge connecting North Korea and China, but instead we ended up in jail, I wasn't convinced that this guy was telling the truth. But no matter what was going to happen, we were helpless. Everything seemed out of our control, and all we could do was wait and hope.

"How do you like the food?" the other official asked. "It was prepared by our soldiers. The vegetables are local to this region. We do not use any pesticides." Then he added, "It's organic."

"It's really good," we responded truthfully.

After lunch, two new escorts arrived to take us to Pyongyang. These officials laid out our belongings, along with a list describing each item. As before, they seemed to be obsessed with totaling up the money and making sure that every cent was accounted for.

We got into an SUV with one of the escorts sitting in the backseat with us. I was by the window, with Euna in the center, and the escort by the other window. We were told to look down and close our eyes. Euna and I clasped hands the entire way. The mountainous roads were steep and rocky. I had to be careful that I didn't knock my wounded head up against the window as we curved through the rough terrain. But it wasn't easy, and I found myself banging up against the glass several times, causing my head to pulsate. I was able to look up occasionally and saw nothing out the window other than dry, barren fields. We rarely, if ever, passed other vehicles on the road. After about six or seven hours, it began to get dark, and we pulled into a gloomy motel. It seemed we were again being handed off to new escorts, who met us at this location. In a dark room lit by candles because there was no electricity, the new people in charge took note of our belongings, once again carefully counting the money.

Dinner was brought to the room. One of the escorts looked on as we ate a simple meal of rice, soup, and vegetables. He seemed almost apologetic about the lack of meat, and said in Korean, "I hope you are okay with not having meat. It's become very popular for people to eat only vegetables, because it's healthier. These vegetables are all organic. There are no pesticides."

This was the second time I had heard someone express pride in the pesticide-free vegetables. I knew they were just covering up for their country's shortages. I was embarrassed that our being Americans had caused such a reaction.

"The food is delicious," I responded with a gentle smile, asking Euna to translate. "We have a lot of vegetarians in the United States too."

After dinner, Euna and I were placed in separate rooms. We were told that under no circumstances were we to open our doors, and that someone would come for us in the morning.

### ⸎ LISA

LINDA MCFADYEN WAS ALSO assigned to our case from the State Department's Office of Citizen Services. She was in contact with both Euna's and our family every day. More than anything she became our friend and the shoulder my mother cried on.

Mom called Linda every morning, as soon as she woke up, to find out if there were any updates. Most days Linda had nothing to report, but she was always patient and gracious with her time. I know this was hard to do because, for lack of a better way of phrasing it, Mom got crazy. She walked around the house like a zombie and she looked like one too. Her hair stuck straight up from not washing it, and she wore the same brown sweater and gray sweatpants for days on end. She had already been diagnosed as an insomniac, and her prescription sleeping aids weren't working. A family friend gave her some enhanced baked goods that were moderately successful in helping to relax her and get her through the tear-filled days and nights. Two full weeks into our ordeal, Iain finally had to tell her to shower and change her clothes. She became so frantic that every time I left her house, I would get nothing short of ten calls from her asking if I had heard anything new. In frustration, I often exclaimed, "No!"

Dad has always been known as the funny guy with the lewd punch lines, so for the most part, he held himself together when he was out in public. Privately at night, however, he would call me from his home in Sacramento and break down. It's difficult to describe how hard it was for me to hear my tough-guy father crying so deeply and painfully.

"I miss her," he would say. "I miss my little girl."

Kurt Tong and Linda McFadyen started scheduling a weekly conference call with our family and Euna's husband, Michael, on Fridays. Most of the time there was little to report. Kurt, Linda, and others who were in the room at the State Department during the calls were often on the receiving end of my foul-mouthed tirades, which erupted out of frustration.

Former Vice President Gore became a main point of contact with the highest-level people in the State Department and the White House, and without him the morass of layers of government would have seemed impenetrable. Though I had connections within different branches of government, it was comforting to have someone with such intimate knowledge of Washington's inner workings.

Gore was relentless from the start. He told me of his close friendship with Deputy Secretary of State Jim Steinberg, who was the principal deputy to the secretary of state. In other words, he was the person who had her ear. Gore said he was speaking with Steinberg regularly, and it was determined early on that the best way to handle the situation was to get China to help persuade the North Koreans to let the girls go. China was an ally of North Korea, and North Korea relied on it for its economic survival. Within days, Gore had reached out to a number of high-ranking Chinese officials, including the Chinese ambassador to the United States, Zhou Wenzhong. From early conversations, it seemed as if China was game to help.

But I wasn't content to stop there, and I started furiously looking for anyone with knowledge of North Korea. A few names came up repeatedly. One was Bill Richardson, the governor of New Mexico. He has had more successes than any other American in negotiating the release of Americans from North Korea. From his days as a member of Congress, Governor Richardson has been active in U.S.–North Korean affairs. In 1996 he helped rescue Evan Hunziker, a young man who was accused of, but not tried for, espionage when he swam naked from the Chinese border into North Korea. Plagued by years of mental illness, Hunziker committed suicide a month after his return to the United States. Most recently, in April 2007, Governor Richardson made his sixth trip to the Communist country to recover the remains of American servicemen killed in the Korean War. More than thirty-three thousand American troops died in the Korean War from 1950 to 1953, and more than eighty-one hundred are still listed as missing.

I had never met Governor Richardson, but I was initially

hesitant to reach out to him because of the controversy that had surrounded him in the weeks after Barack Obama won the presidential election. The president-elect offered him the position of commerce secretary, but Governor Richardson abruptly withdrew his nomination when an investigation into some questionable business dealings in his home state arose. He would be exonerated in August 2009, but this was still March, and I wasn't sure if the Obama administration would welcome the governor's participation.

Making things even more complicated, when Richardson dropped out of the 2008 presidential race, he endorsed Barack Obama instead of Hillary Clinton. According to press reports, this upset former President Bill Clinton, a longtime friend of Richardson's, who first appointed him ambassador to the United Nations and later secretary of energy in his administration. Now Hillary Clinton was the secretary of state. It seemed that, by all accounts, Richardson had fallen out of favor with the powers that be.

Still, a friend gave me the governor's contact information, and he responded to my call immediately. He sounded like someone who was unsure of how he stood with the administration, but he said he would try to help us as a private citizen. I liked Governor Richardson from the start. He seemed more like a regular guy than any other politician I'd ever met.

He asked me if the State Department had a plan for how to deal with our situation, and I told him that Beijing was being solicited for assistance. "The North Koreans hate dealing with China!" he tersely warned. "Trust me, the North Koreans will become very upset if the U.S. tries to involve China in any way."

He went on to say something that would be repeated to me by a number of ardent North Korea watchers: "What they [the North Koreans] want is to deal directly with the United States. North Korea is insulted by the six-party talks."

Begun in August 2003, the six-party talks are a negotiating process involving six nations—North Korea, the United States, South Korea, Japan, China, and Russia—to bring about a peaceful resolution to the security concerns provoked by North Korea's nuclear ambitions.

Governor Richardson promised to call the State Department to offer his perspective on China's involvement and what he believed North Korea wanted. He also said he would reach out to his contact in North Korea's Foreign Ministry to see if he could get any information about our case.

Several days after our first conversation and just over a week after Laura and Euna's initial detainment, Richardson told me that both President Obama and Secretary Clinton had officially asked him to take on our case. They also asked him to maintain a low profile, given the sensitivities among the countries that neighbor North Korea.

"I'm telling you that they called me because I have already been advising you," the governor explained. "You can't say anything to anyone. Okay?" He went on to say that he'd told the State Department to cut China out of the process.

Governor Richardson felt confident that he could secure the girls' release. He said he was going to begin making overtures to his North Korean contacts and let them know that he would be point man for any negotiations should they happen. He also made sure to note that he was going to tell them he was prepared to jump on a plane to Pyongyang immediately. It gave us some relief to know that Governor Richardson was working on our behalf and that he still maintained his North Korean connections. At the end of our conversation, he reminded me that he had never failed to bring people home from North Korea and he had no reason to believe that this time would be any different.

# CHAPTER THREE

# going to pyongyang

∿↝ **LAURA**

THE NEXT MORNING, WHILE we were still at the motel, a
doctor came to my room to check and sanitize my wound. As
he removed the bandage, the room went dark. Another power
outage. An official moved to the window to let in some light, and as
he pulled the curtain aside, the entire rod of black fabric came crashing down and natural light flooded the room, creating a hazy glow.

After the doctor finished his work, I was left in the room alone.
I sat and peered out the window. About fifteen yards away, I could
see an electric train filled with downcast commuters. Also, there appeared to be a factory off in the distance and row upon row of small,
ramshackle houses.

A few hours later we set off again on the dusty ride to Pyongyang. Euna and I clutched each other's hands in fear and kept our
eyes closed. Early on, while traveling up a mountain road, the car's

engine began to fail. We pulled off to the side, and when the escorts left the car, I lifted my head and looked around. People were walking and riding their bikes, but it was eerily quiet. People didn't seem to be talking to one another, just going along their way. Finally, we were on the road again.

By nightfall, we entered a small town and pulled up to a dilapidated gray building. The streets were dark and empty. The dim glow of candlelight shone through some windows of what looked to be a three-story apartment complex or dormitory; the others were pitch-black. There was no electricity. We were led into what looked like a family's kitchen and living room. The only decorations on the walls were portraits of North Korea's current leader, Kim Jong Il, and his father, the previous leader of North Korea since its founding in 1948, Kim Il Sung. It was hard to tell if this was a home that doubled as a restaurant or if our escorts had arranged for the woman of the house to prepare a meal for us. The room was sparse, with just a low folding table in the center. We were each given a plastic box of rice, an egg, and kimchi, along with a bowl of seafood soup. A skinny kitten roamed around the room and rubbed up against our feet, waiting for scraps.

As I picked out the bones from the fish in the soup, the officials commented on how well I was able to use chopsticks with my left hand. Every time I've gone to Asia, people have commented about my being left-handed. In some Asian cultures, left-handedness is not desirable, and many parents force their children to use their right hand instead. Tonight I wondered if being left-handed made me seem even more peculiar to the authorities. Here I was—a left-handed, black-and-blue-bruised American of Chinese ancestry who could barely speak Chinese. I felt relieved that I at least had Euna to help communicate.

After the meal, we drove for roughly twelve more hours before pulling up to what appeared to be a government building. Another

SUV was waiting in the parking lot, and it was clear that meant they were separating Euna and me. Suddenly my mind was swirling with questions. Would we ever see each other again? Were they going to let one of us go?

At this point Euna and I were allowed to use the restroom in the government office before beginning the next leg of the journey. We squeezed each other tightly and told each other to be strong. Without Euna, I was alone and lost.

I was taken to the new car, squashed between three officials in the backseat, and then we were off. I tried to make myself seem as small as possible and not take up too much room. Periodically a man in the front seat shone a flashlight onto my face to make sure I was looking down and my eyes were closed. I shuddered in fear each time the sharp rays penetrated my eyelids.

The distance from the border region where we were apprehended to the capital city is roughly three hundred and fifty miles. Our journey, over winding, rocky mountain paths and narrow dirt roads, ended up taking around twenty-four hours over the course of two days.

All of a sudden I could feel that we were driving on paved roads for the first time since we began the trip. My eyes had been shut for the past five hours. Finally I could sense the vehicle go up a steep driveway and heard a guard dog bark ferociously.

When I stepped out of the car and opened my eyes, I saw that it was early morning and could feel the raw wintry air. It didn't get any warmer when I entered the two-story building. A large chandelier hung in the entryway illuminating a giant mural of Chairman Kim Jong Il walking through a park on a brisk autumn day. Kim's austere image and that of his father are constants throughout the country, but especially in the capital.

As I looked at his iconic image on the wall, I wondered if Kim knew we had been taken captive in his country. I had known before our apprehension that there was a lot of speculation about Kim's deteriorating health after the stroke he suffered in 2008. Now it was unclear if Kim was the one calling the shots or if his ruling generals would dictate our fate. I didn't know which would be worse.

I was led into a corner room with portraits of the father and son hanging high on a wall. A stern-looking official with the face of an aging bulldog rose from behind a desk and started looking me over from head to toe. I bowed toward him politely. Two young women who appeared to be in their midtwenties followed his lead and leaped up from a couch in the corner to peer at me with ice-cold stares. The man led me into an adjoining bedroom and began speaking to another official who gruffly translated his orders to me in Chinese. I listened intently, hoping my basic knowledge of Mandarin would be enough.

"You will stay here and rest," he instructed. "Use the bathroom to wash up. If you need anything, ask one of these girls. Do you understand?"

"Yes, thank you," I replied in Mandarin. The two men left the room, leaving the two women guards to watch over me.

In the bathroom, I looked in the mirror for the first time in days. I could barely recognize myself. I was pale and gaunt, my right eye was black-and-blue, and my jaw was still swollen. A bandage was wrapped around my head covering the gash. I peered down at the neck of my turtleneck sweater and began picking off bits of dried blood that were stuck there.

"Who are you?" I whispered to myself. "How could you let this happen?"

It had been nearly a week since I'd spoken with Iain. And about ten days since my last Skype conversation with Lisa. Rarely would a day go by without my speaking to at least one of them. I'd never felt more alone or confused.

Iain and I met twelve years ago when we were both in college.

We had mutual friends who were all going to a concert, and a group of us gathered at Iain's apartment before heading to the Shrine Auditorium in downtown Los Angeles. I literally bounced into Iain while dancing to the electronic beat of the Chemical Brothers, the British duo that helped popularize the electronic dance movement in the 1990s. While we didn't say much to each other in the blaring hall, I felt an instant connection between us. The next day I decided to get hold of Iain's telephone number and, at the urging of my roommates, rang him up. When he answered, I was at a loss for words. "Um, hi. This is Laura. I met you last night at the concert. I think I left my driver's license at your apartment," I said nervously, all the while holding my license in my hand. We ended up talking for an hour. A week later, Iain called me and asked me out to dinner and a movie. It wasn't until our one-year anniversary that I told him I'd had my license all along.

Iain was my first serious boyfriend and the only person I felt important enough to introduce to my family. I wanted him to meet Lisa. She was my best friend, and her opinion meant more to me than anyone else's. I was afraid that she might disapprove of our age difference—Iain was ten years older than me and working on his second master's degree. Anytime I mentioned him, she skeptically brought up his age, even though she had dated plenty of older guys. But I knew she was just being my protective big sister. Lisa and I consulted each other about practically everything. If I didn't like someone she was dating, she usually ended the relationship soon after. I didn't want that to happen with Iain. Fortunately, she and Iain got along from the start. Over the years they'd become like brother and sister. Iain, who is Australian-British, doesn't have any family in the United States and my family quickly adopted him as their own. At this confusing, scary time, there was some consolation in knowing that he would be with my family and not alone.

I made my way to the bed, sat up against the wall, and hugged my legs against my chest to provide extra warmth. It was so cold in the room that I could see my breath. But the blanket beneath me felt unusually warm. I reached my hand inside the yellow comforter and felt the heat from an electric blanket. I scrambled to get under the covers. Just as I was getting warmed up, the electricity went out. I wrapped the blanket around me tightly, not wanting to let any of the heat get away.

Compared with the cell I'd been in before at the border, this room felt spacious. I was grateful to be lying in a bed and to have an adjoining bathroom. And though the guards watching my every move seemed cold and intimidating, I was thankful to have some human presence.

I poked my head out from beneath the covers and shifted my eyes over to the guards' area. Their room was connected to mine by a pair of folding doors, which were always left open so that the guards could look in at any time. Their quarters contained two couches, a desk, a small coffee table, a bookshelf with a collection of Communist teachings, and a television set. There was also a tall, freestanding heater and air-conditioning unit, which was not working.

The two guards were sitting and reading quietly. I hadn't encountered or seen anyone like them in North Korea. There was a casualness to their attire. They both had on makeup and appeared to be well groomed. Min-Jin, the older guard, who seemed to be the one in charge, appeared to be in her late twenties. She wore black slacks, black heels, a red turtleneck sweater, and a puffy black trench coat with faux fur trim around the hood. Her hair was pulled back neatly in a ponytail. Her red sweater was the first clothing I'd seen that wasn't a dark shade or army green. Kyung-Hee, the younger guard, had a round baby face. She had short straight hair and wore black slacks, sneakers, and a light-colored jacket. Every now and again I sensed her looking over at me curiously. I decided I'd try to communicate with them.

I crawled out of bed and took a few steps toward the room where they were seated. I tried to remember the Korean phrases Euna had taught me.

"Good morning," I said, stuttering as I tried to put the words together. "I'm sorry, I don't speak Korean. Do you speak English?"

"What do you want?" Min-Jin replied in broken English and in an arrogant tone.

"Oh, you can speak English," I responded, smiling in relief. "It's nice to meet you. Do you know what is happening with me? Is someone going to come see me?"

"Just wait," she replied harshly.

Kyung-Hee shot me a stony glare. I went back to the bed and buried myself inside the covers.

I was awakened several times during the night by the dog that was growling outside. I couldn't make out any other sounds. I didn't know if I was in the capital city, in the outskirts, or somewhere else entirely. There were windows in both rooms, but the curtains remained closed. I was told not to step anywhere near the windows.

The next morning I was brought into the guards' area and told to remain standing while several authorities crowded into the room. A photographer and videographer set up their equipment and began taking pictures. The bulldog-faced official I'd met the day before entered the room and sat at the desk. In a booming voice, he began reading from a document in Korean. A man standing to my right translated what he was saying. But he wasn't translating into English—instead he was speaking in Mandarin. I interrupted him and explained that I didn't understand what he was saying. This seemed to confuse everyone in the room. Whatever official announcement they were making was not going as planned. I stared at the floor, nervously waiting to be told what he was saying. I could sense from his tone that he wasn't going to tell me I'd just won a ticket back home.

The photographer kept snapping pictures of me, while various authorities consulted with one another in hushed voices. Finally they called over Min-Jin, the female guard who spoke some English. Until this point, she'd seemed confident, and had an almost superior air. But now that she was being asked to perform, she suddenly became scared and vulnerable.

Once again the official began his speech, with Min-Jin translating. He said that journalists have a duty to uphold the truth, to report on stories about injustice. Then he said: "You were trying to distort the truth and spread falsehoods about the Democratic People's Republic of Korea [DPRK]. You are not a good journalist."

I was surprised that his denouncement was devoted to my job as a journalist. He actually said very little about stepping into North Korean territory. I wondered how much they knew about the report we'd been working on.

"I'm very, very sorry," I responded, in tears.

"If you confess your crimes, openly and frankly, and express regret for your actions, there may be forgiveness," the man continued. "However, if you are not honest and frank, there will be punishment. Do you understand?"

"Yes," I responded dejectedly.

"Speak up!" the man ordered.

"Yes, sir!" I loudly proclaimed. "I understand."

Then they all filed out of the room, and I was left there with the two female guards.

"Go back to your room," Min-Jin instructed. "Someone will come to see you later."

"Thank you for translating," I said. "You did a very good job." Her lips curved into a slight smile. I could tell she appreciated the compliment.

I was relieved that I hadn't been given my death sentence and decided that the mention of forgiveness was a positive sign. But the word *punishment* echoed in my mind. What sort of punishment could I be facing? I hoped a confession and an apology would be enough to win back my freedom.

A few hours later I was ordered back into the guards' room. A man entered carrying a red notebook. He introduced himself as my investigator, Mr. Yee. I bowed to him respectfully, and he motioned for me to sit down. He was dressed in the standard black suit, with a pin showing the founding leader, Kim Il Sung, on his chest. His hair was well groomed, and he smelled of fresh soap. He looked to be in his late forties. He had a kindness in his eyes that contradicted his stern demeanor.

He sat down at the desk, lit up a cigarette, and let out a few puffs while looking me over. Using Min-Jin to translate, he explained, "I will be handling your investigation. I will visit you every day, and I expect you to cooperate fully with my investigation. I am in charge of you, so anything you need or any questions you have, you can ask me."

Min-Jin's English was so elementary that I had to ask her to repeat herself several times before I fully understood what the investigator was saying. Mr. Yee handed me a few pieces of paper and told me to write down my education, career, and family history, with the ages of my immediate and extended family, including aunts, uncles, cousins, and their spouses. He exited the room, leaving Min-Jin to instruct me.

I was confronted again with the dilemma about describing what Lisa did for a living. I figured some government workers might have Internet access here in Pyongyang and that they could easily discover the work Lisa had done in North Korea. I kept remembering the words from that morning—*forgiveness* and *punishment*—so I decided to be honest. Next to Lisa's name, I wrote: "Freelance Correspondent." However, I did not disclose that Lisa was currently working for *The Oprah Winfrey Show,* or that previously she had worked for National Geographic Television, CNN, and *The View*.

I was also worried about saying what my father's job had been before he retired. My dad was once a civil service employee for the U.S. Air Force at McClellan Air Force Base in Sacramento. He'd been sent to Vietnam during the war to conduct safety of flight inspections on aircraft. I didn't want to tie my family to the U.S. military in any way, thinking this might further incense the North Korean authorities. "Retired," I wrote next to his name. I purposely misspelled it "McClelan," leaving out an *l,* and I didn't say it was an air force base.

Mr. Yee returned, this time followed by another man, Mr. Baek, who introduced himself as a university professor who had been assigned to be the investigation's official translator. Mr. Baek wore a short-sleeved, checked shirt and slacks. I liked him the minute I saw him. He had a round, friendly face with spiky salt-and-pepper hair. It was the first time I'd seen a man wearing anything but the traditional Communist garb. His English was impeccable. Mr. Yee looked over the papers I had just filled out about my family and occupation.

"You are a journalist working for Current TV, is that correct?" he asked speaking through Mr. Baek.

"Yes, sir."

"Your husband. What does he do?"

"He works in the finance industry. He manages money."

"Can you be more specific?"

I then went into a detailed explanation of Iain's profession, something in which Mr. Yee seemed to take great interest considering that the free market does not exist in North Korean society. When translating, Mr. Baek mimicked Mr. Yee's tone with such precision that it felt as if I was speaking directly to the official without the aid of another person.

When Mr. Yee left the room for a brief moment, I felt comfortable enough to start talking to the translator, Mr. Baek.

"Sir, do you know what is happening, what this process will be like?" I inquired.

"I don't know a thing," he replied. "I was just informed today that I would be translating for this investigation. Beyond that, I don't know a thing."

"Do you know why I'm here? Do you know what I've done?" I was trying to see if the authorities knew more about the story Euna, Mitch, and I were working on than we'd said.

"I haven't a clue," he answered.

"You speak English so well," I added. "Are you from somewhere else, like Singapore?" I thought I'd picked up on a slight Singaporean accent.

"I'm from North Korea," he replied, smiling. "I work in the international affairs department at the university. It's my job to speak English well."

"Well, I'm very glad to meet you," I said. "It has been so hard to communicate with anyone here, and it's a relief to speak with someone who I know understands what I'm saying." He acknowledged my compliment by nodding, but his face was totally expressionless.

The interrogator, Mr. Yee, returned and started questioning me about what we had been doing before our arrest. I began with the same explanation I'd rehearsed—that we were working on a story about the border region. Before I could finish my sentence, Mr. Yee jumped in.

"We know you were working on a story about defectors," he said. "Let me add that this is an investigation, and if you are not completely frank with us, you could face the worst consequences!"

I quivered in fear, while at the same time I was utterly flustered.

Had Euna already confessed about our documentary? Were the North Koreans getting intelligence from the Chinese, who had perhaps detained Mitch? Regardless of where they were getting their information, I knew what I had to say.

"Yes, we were working on a story about defectors," I began. "We'd interviewed various people who had left North Korea about why they left."

"Why did you cross the border into North Korea?" Mr. Yee asked.

"We wanted to film on the river to show where defectors are crossing. We never meant to enter your country," I explained. "But the guide we had hired to help us walked across to the other side and motioned for us to follow, which we did. We returned back to China after taking only a few steps onto the soil. We didn't mean any harm. But I know that it was wrong. And I'm very, very sorry."

Mr. Yee took another puff of his cigarette, pressed the butt into an ashtray, and got up from the desk.

"I don't believe you," he said coldly and left the room.

Again I was alone with Mr. Baek. Desperately wanting someone to confide in, I told him how scared I was and then waited to see if he would reply.

"It is a very bad time to be here," he said in perfect English. "Things are quite tense between North Korea and the U.S."

"I truly never planned on crossing the border," I explained.

"Well, I really hope everything works out and that you can go home," he said sympathetically.

"I hope so too," I replied. I felt grateful to have come in contact with Mr. Baek. I was glad he would be serving as my official translator for the investigation.

The day's session ended with Mr. Yee instructing me to write down every aspect of what we had filmed prior to our arrest. He handed me several sheets of paper, and he and Mr. Baek then left the room.

Our PARENTS SHOULD HAVE never been together in the first place. He was a gruff man, happiest when he was fishing or hunting with the guys. She was distractingly beautiful but insecure and didn't have many friends because she had just arrived from Taiwan and knew no one in the United States other than her sister.

Our father, Doug Ling, came from highly educated parents who met in Hong Kong in the 1930s. His father, H.T. Ling, was part of an elite group of Chinese students who were given permission by the government to attend university in the United States in 1920. After receiving an undergraduate degree from New York University and an MBA from the University of Colorado, H.T. was recalled to China to assist in the war effort when the Japanese invaded Manchuria in 1931. Doug's mother, Lien, was the daughter of missionaries in Malaysia. When she was a teenager, she went to live in Hong Kong and earned a degree in classical music from the London School of Music program there. Lien was an unusually well-educated, well-spoken woman with a very strong voice and opinions. This impressed H.T., who was doing his compulsory military service in Hong Kong while Lien was teaching piano there. They married, had two children, and lived in the British colony until they emigrated to the United States in 1948, when Doug was eleven years old.

The Lings ended up in a small suburb outside of Sacramento, California, called Carmichael, where a few of their relatives owned a Chinese restaurant called Sun Ar. They soon learned that the locals weren't exactly friendly toward "Orientals." Even with their impressive degrees and perfectly spoken British English, H.T. and Lien had tremendous difficulty finding jobs that were commensurate with their levels of education. And it didn't help that H.T., who was descended from a long line of Chinese scholars and government officials, was a bit arrogant and refused to do anything beneath his skill set. He spent a great many years, with little success, trying to convince companies in Sacramento to hire a Chinese man for a management-level position.

The responsibility for taking care of the family fell on Lien and young Doug. At first they were unable to get any locals to rent them a home, but eventually they were permitted to live behind someone's main house in a converted chicken coop. With the little money they had, the family took over the restaurant Sun Ar. Lien taught piano in the mornings and cooked in the restaurant at night. Doug worked slavishly in the kitchen, seven days a week. He reported to the restaurant right after school ended, which meant that he never went to a high school football game; he never went to a prom. To avoid being teased by the kids in his blue-collar community, he developed a sharp wit and a foul mouth. Somehow, making *fuck, goddamn,* and *shit* a regular part of his speech allowed him to fit in better, even if it shocked and dismayed his highfalutin parents. He could make his friends laugh, particularly when he'd deride other races, including his own. He laughed when his classmates referred to him as "Chinaman," even though it made him seethe inside.

One day when Lien was on a trip to downtown Sacramento to purchase bean sprouts for the restaurant, she met a factory worker named Mrs. Wang who had just recently come to the United States with her two daughters, who were twenty-two and twenty. The three women were literally fresh off the boat; they didn't speak any English, only Taiwanese and Mandarin. So another factory worker served as interpreter for Mrs. Wang and Lien, who spoke only Cantonese and English. The worker told Lien that Mrs. Wang's younger daughter was quite beautiful and that her mother was pressuring her to marry. At the time, Doug, who was thirty-one years old, was dating a Caucasian woman, which bitterly displeased Lien. She suggested that Mrs. Wang's daughter Mary meet her son, Doug.

This was our mother, Mary. She was the middle of seven children and the loneliest. Her father was uneducated, but for a time he had been one of the richest and most powerful men in Tainan City, Taiwan. His name was Wang, but he was known as Black Dragon and he was one of the leaders of Tainan City's underground. He owned hotels, some of which were fronts for brothels and casinos. These side businesses paid him handsome returns. During the Japanese occupation of Taiwan, his businesses serviced Japanese soldiers, even though he hated them. His authority was such that local politicians running for office had to seek his approval if they had any hope of claiming victory.

When he was twenty-four years old, Wang noticed a shy but beautiful country girl at the spring festival. He couldn't take his eyes off her and sent for someone to recruit her. Rather than putting her to work, he made her his first wife when she was nineteen years old. They eventually had seven children together over the course of twenty years. But only two of their offspring really mattered—the boys. During his marriage to Mrs. Wang, Wang took on two concubines with whom he fathered twelve children. Wang claimed a total of nineteen offspring, but it was rumored that many more children bore his genes. Mary's only contact with her father was when he came to their house, patted the girls on their heads, and gave them a few dollars. He spent most of his time with his other wives and the other women he paid. Mary's mother grew to hate her husband and warned little Mary to be cautious of men. Mrs. Wang never knew what love felt like, and her fourth daughter would never know it either. Her mother's words always stayed with her—never trust men.

Mrs. Wang ran one of Black Dragon's legitimate hotels.

Wang's second wife ran the ones in which brothels operated. From time to time, Mrs. Wang would have to drop things off at the second wife's hotel. The things she saw there disgusted her, and she grew increasingly resentful as the years went by. She tried to shield her girls from the scenes of debauchery that surrounded them, but it was no use. On a few occasions Mary was sent to take things to the different businesses. Listless and tired workingwomen frequented some of the hotels. Young Mary once saw a woman with red lips crying in a room. She wanted to help the woman, but she was only a little girl and her mother told her to stay away from "them."

Later that day, police officers and an ambulance arrived. The woman that Mary saw crying had hanged herself. The ambulance workers carried out the body, which was covered in a sheet. The only things visible were the woman's high-heeled shoes. Mary never forgot what she saw. It made her heart hurt. Mrs. Wang began to stash away large sums of money in places where no one else could find it. She knew that one day she would find a way to take her children, leave, and never come back.

The year Mary finished high school, her father lost his entire fortune to gambling and could no longer support his families. Mrs. Wang saw this as an opportunity to leave. Her eldest daughter, Jeanie, had gone to the United States and married an American citizen. Mrs. Wang temporarily left her three youngest children with an aunt and took Mary and her sister Ruth to live with Jeanie and her husband in Sacramento.

Mary and Doug's first meeting was pretty awkward, considering that they knew their parents had as much as married them off. But he did find her attractive, albeit too skinny. They could not have been more different. Doug was content with a simple existence of fishing with the boys and eating his meals in front of the TV. Mary was ambitious. She loved nice clothes and wanted to see the world. On top of it all, they could barely communicate. Mary had difficulty with English, so for the most part they just smiled and exchanged polite gestures.

Doug was on his best behavior, but that didn't last long. He was ten years older, and he'd already had a lot of experience going to bars. On their first dates, he brought Mary along, but she hated the scene and didn't like Doug's friends. She thought they were crass and rude. She didn't drink; he drank a lot. They started going to movies where they didn't have to talk much. Four months later, on March 8, 1969, they were married.

Their first year of marriage was decent enough. Doug had been working as a supervisor at McClellan Air Force Base in Sacramento, and Mary went to school to learn English while working on the weekends as a waitress at a local Chinese restaurant.

They hardly saw each other. Their inability to fully understand each other prevented them from developing a real intimacy. Even so, he came to love her. She, on the other hand, could never forget her mother's words to her, "Never trust men." She couldn't love.

Life was just perfunctory: they worked all day, and at night when they were home, they watched TV. This went on for years, and eventually Doug started going out more with the guys, hitting up the bars a couple of nights a week. He came home smelling of beer, talking loudly, and rambling on about how much he loved his wife. Then he would stumble around until he eventually passed out in bed. This made Mary feel lonelier than ever, so she decided it was time to bring a new friend into the world. On August 30, 1973, I was born.

I was only two years old when my mother's belly started growing bigger and bigger. I recall Mom telling me that my baby brother or sister was inside of her, and the thought of it sent such a shock through me that I still remember it to this day. It was amazing. I kissed my mom's tummy and talked to it all the time. I knew it was a girl; it just had to be a girl—that's how much I wanted a sister. On December 1, 1976, Laura came into my life.

From the day she arrived, I remember feeling like we were a team. I was glad I wouldn't have to deal with the family drama by myself, even if Laura was a baby for much of it. Just knowing I had a partner made things a bit easier to bear, especially because our parents fought throughout our childhood.

Mom wanted to do things like go to the movies and try new and different restaurants. Dad hated movies and would only eat in the same Chinese restaurants. Plus, Dad worked all the time, and on the nights when he'd come home early, he'd crack open a beer or two and sit in front of the TV for the rest of the night. Mom seemed detached and depressed a lot. But none of this meant they were bad parents. Despite the tumult in their relationship, both of our parents were nothing but loving to us all of our lives. Some of my fondest memories include the nights when, with her melodious voice, Mom would sing us to sleep. She always sang the same song, "Edelweiss," from the musical *The Sound of Music*. Of course, because of her difficulty with English, it always came out "A dell voice." To this day, Laura and I still sing it that way.

On the nights that he worked late, Dad would return home and tiptoe into our room thinking we were asleep. He would lean down and give both of us little pecks on the forehead and whisper, "Daddy loves you." I loved the way his prickly mustache felt on my skin; I loved how much he loved us.

I was seven years old and Laura was four when our mother and father finally decided to go their separate ways. We stayed in Sacramento with our father because our parents didn't want to uproot us from our school and community. Plus, we had Grandma. Dad's mother, Lien, lived with us until she was struck by senility when we were teenagers; she was our source of stability. Grandma was a self-assertive woman, particularly when it came to her faith. She always felt the need to "save" people, which sometimes embarrassed us. It was hard enough to be among the only Asian kids in our community, but Grandma would make the neighborhood kids sit through obligatory Bible study whenever they'd come over. She would make us memorize verses and quiz us from time to time on what they meant. Needless to say, our house wasn't the most popular destination for kids.

Grandma thought Halloween was a pagan holiday, so every year on October 31, she would turn off the lights in front of the house and make Laura and me sing church hymns at the top of our lungs to drown out any doorbell sounds. We were the only kids who came to school the next day with no candy. Even so, other than each other, our grandmother was the most important person in the world to us. She taught Laura and me to respect ourselves and to become highly self-sufficient. Most important, she taught us how to be strong women.

We were everything to her. The two things that Grandma loved most were God and her granddaughters. When we were teenagers, the nice people at the nursing home would tell us that the only names our grandmother ever brought up were Laura's and mine.

"Where's Lisa and Laura?" she would ask each morning. "Are Lisa and Laura coming to see me today?"

We lost her in 1991.

Our mom moved to L.A. shortly after the divorce. She started working as an office manager in a law firm and moved up the corporate ladder quickly. We saw her fairly often. She would fly to Sacramento at least once a month and stay for three days to a week at her sister's home. During the hours when Dad was at work, Mom would hang out with us in the house. We loved when she would bring us cool clothes from L.A., and all the kids envied our stylish new wardrobe. The clothes masked a lot of sadness. We would have given it all up in a second to have both of our parents together and happy.

Laura and I became seasoned travelers at very early ages because Dad would put us on a plane to spend the summers with our mom down south. A lot of the flight attendants came to know us; we were two little girls on a very big plane. Having each other made the hard parts so much easier.

Despite our challenging beginnings, Laura and I remained very close to both of our parents. They were both victims of circumstances that were larger than they were, and Laura and I have always been proud of the way they tackled so many of the issues that confronted them as young people. In the end, our parents' divorce may have been the best thing to happen to us because they were so unhappy with each other. But when we were going through it, it was a nightmare. I think Laura was too young to remember most of the dark and ugly episodes that occurred when we were kids, but I wasn't. Holes in the walls and broken objects litter many of my childhood memories. As the sounds of yelling and shrieking cries were coming from downstairs, I would look over at the baby girl and think—I just want to protect her.

Now we were all grown up, and she was in trouble.

∾∾ **LAURA**

O VER THE NEXT DAYS and nights, I fell into what seemed like a black hole that included daily interrogations, psychological intimidation, and virtual isolation. Some nights after I'd been grilled all day about my past jobs and other assignments, I curled up into a ball in a dark corner of my room and sobbed profusely, wishing I could make myself small enough to just disappear. I feared I might never see my sister, my parents, and Iain ever again. I hated myself for putting my family through such pain.

For a few days, I was worried that I might actually be pregnant.

Although I couldn't bear the thought of giving birth and raising a child in a North Korean prison, a part of me hoped I had a baby inside me. It made me feel less alone thinking that I might be carrying a child. I also thought that being pregnant might cause the North Korean authorities to be more sympathetic to my situation. *Maybe this child is a gift from God,* I thought. *Perhaps I was meant to get pregnant here so the baby could save Euna and me by giving the authorities the compassion to release us.*

Days later, whatever fears or fantasies I had were put to rest. Any chance of being pregnant was gone. I was both relieved and demoralized. It crushed my heart to think I might never get to start a family with Iain. Though he was anxious to have a child, Iain had been patient with me while I pushed aside the idea of having a baby and instead focused on my career. Now, I thought, the chance might be gone forever.

I tried my best to keep my mind from wandering and longing for my family. I knew I needed to concentrate on the investigation so I could satisfy the authorities with my answers. I also wanted to be careful not to endanger any of our defector contacts or get myself in any more trouble than I already was.

During one interrogation session, Mr. Yee started asking me about my or my company's ties to the U.S. government.

"Al Gore is your chairman," he began. "So is your company connected to the government?"

"No, not at all," I answered. "Vice President Gore wanted to start this company so that all people would have a voice in the media, not just the big corporations. It's a network where anyone can participate. For example, if you disagree with what's happening in North Korea, you can comment about it on our Web site. And if you agree and support the North Korean government, you can voice your opinion too."

"Then who is funding your project here? Are you receiving

money from the U.S. government to produce this documentary?" he asked.

I knew what he was asking and was afraid he might think that the company or I was being bankrolled by the CIA. I tried hard to convince him there was no connection whatsoever to the U.S. government. I went into a detailed explanation of the U.S. cable business, the media conglomerates, and advertising. He seemed genuinely intrigued and often interrupted me to ask questions about how advertisements work and about the various tiers of cable packages that consumers can choose from. Still, the question of whether or not the U.S. government was involved in our documentary would continue to rear its head multiple times throughout the investigation.

EVERY FEW DAYS a doctor and nurse visited me and cleaned my head wound. The doctor, a slim man with a nervous twitch, often gave off loud sighs as he inspected the area, causing me to worry. But each time, he assured me there was nothing to be alarmed about. The nurse transferred a cotton pad soaked in alcohol to his metal tong, and he rubbed that around the wound. Every time the alcohol touched the affected area, it would feel like a thousand needles sticking me. Rather than stitching the gash, the doctor preferred to let it close up on its own.

Since the beating, I had been having frequent headaches. When Mr. Yee asked me how I was feeling, I told him about the shooting pains in my upper left lobe.

"You're young. You will get better over time," he replied. "I wish you hadn't resisted when the soldiers apprehended you. You know what happens when you resist arrest in the United States, don't you? It goes on your police report."

All I could do was nod in agreement, even though I was fuming inside. There was nothing to be gained from showing him my fury. Still, I didn't want him to report that I had opposed the soldiers on the border. I stopped myself from blurting out, "In the United States, police brutality is a crime!" But I couldn't just sit there and accept this lie. "I did not resist, sir," I said defiantly. "Do you really think a girl of my size would challenge a fierce soldier with a rifle?"

Mr. Yee never accused me of resisting arrest again, and I refrained from complaining about my throbbing head.

Each day these rigorous interrogation sessions became a delicate balancing act. I tried to answer the questions while being careful to avoid revealing information that might endanger our sources or the interview subjects who had opened up their lives to us. Early on in the questioning, I told Mr. Yee that our team had received guidance from a pastor who worked in Seoul. Pastor Chun Ki-Won and his network, the Durihana Mission, have helped hundreds of North Korean defectors escape from northern China via a so-called underground railroad that takes defectors through treacherous terrain in countries including Laos and Thailand. Once inside these countries, defectors apply for amnesty through the South Korean diplomatic mission. If their request is granted, they are flown to Seoul to begin new lives.

The North Korean regime sees Chun as an enemy. I knew I needed to confess our connection to him because he was such a central figure in our project. But I feigned having no recollection of his name or that of his missionary group. I could tell this was frustrating to Mr. Yee, who kept asking me how I could forget the name of someone with whom we'd worked so closely.

"How can you call yourself a journalist and not even know the name of someone you interviewed?" he asked.

"Korean names are very difficult for me to remember," I said. "That's why I just called him Pastor the whole time."

Mr. Yee questioned me about Chun for two full days, and as

time went on, he became increasingly annoyed that I would not divulge his identity.

"We already know his name," he finally said on the second day. "His name is Chun Ki-Won."

I tried to hide any sign of acknowledgment. Irritated, he asked, "Do you want to still deny that Chun is the person who helped you? Might I add that when you are speaking to me, you are not speaking to me as your investigator, you are speaking to the DPRK law!"

"I honestly don't remember his name," I said. "If that's what you say his name is, then I'm sure you must know what you're talking about. All I know is that I called him Pastor, and since I am speaking to DPRK law, I don't want to lie."

Mr. Yee then put forward a question I had been dreading. "Have you ever been to North Korea before?" he asked.

I assumed it wouldn't be long before they learned about a visit I had made to Pyongyang in 2002 while working for Channel One News. I, along with Mitch Koss and a Korean-American correspondent, applied for tourist visas during the North's Arirang Festival, a gymnastics and artistic production celebrating the late leader Kim Il Sung. The show features thousands of children, each holding up and flipping colored cards in unison to create massive mosaic pictures. In one instance, the entire side of North Korea's May Day Stadium might show a scene with North Korean soldiers standing victorious, but with a quick flip of thousands of cards, the picture might change to a gleaming portrait of Kim Il Sung. All the while, tens of thousands of dancers perform intricate routines set to patriotic music on the stadium grounds. Along with attending the highly choreographed performance, our tour group was taken to some of the main monuments, statues, museums, and sights that are the country's pride. During the five-day tour, we were never allowed to roam around alone; even walking across the street from our hotel was forbidden.

"Yes, I have been to Pyongyang," I answered despondently. I explained that we were working for an educational program and wanted to introduce Channel One's student audience to the culture of North Korea.

"So you lied and came here as a tourist, not as a journalist?" he asked.

"We knew we wouldn't be admitted as journalists. I know it doesn't make it right, but I am aware of many journalists who have applied to come here posing as tourists," I explained. "You have such an impressive intelligence network here," I said, trying to be flattering, but also trying to make a point about how lax their security was. "You must know that journalists apply for these tours." I could tell that he was slightly flummoxed by my comment.

Just then, Min-Jin came into the room and asked what size clothes I wear. She said she was going to get me some underwear. I told her I wear a small, perhaps even extra small, and that made her look me up and down with a skeptical eye. The thought of them buying me clothes made me feel ill, particularly because it seemed to indicate that I would not be leaving North Korea anytime soon. Then Mr. Yee reinforced my fears.

"The guard is going to get you some underwear because this is where you may be staying for a while," he said. "How many pairs do you think she should buy? If you do not cooperate with me, I may need to tell her to buy you ten pairs because you may be here for a very long time!"

Later that evening, Min-Jin came into my room carrying some underwear. Recalling Mr. Yee's threat to keep me in North Korea, I was relieved when she handed me just two pairs. I looked at the tag and saw they were size large. But when I tried them on, they were a little snug on my thin frame. I then realized why the guard had given me a skeptical look earlier when I told her my size. My measurements might be small by Western standards, but compared with the average North Korean, I am enormous. I was also given some long cotton leggings and a top to wear as pajamas. I wore these underneath my clothing throughout the day to help keep warm. Although receiving these items seemed to be a sign that I was going to be held captive for a while longer, I was happy to have some clean garments.

I was also grateful to be given three meals a day. Normally, each meal consisted of a bowl of rice, a small plate of vegetables, and a little piece of fried or steamed fish. Sometimes they brought me a bowl of noodles, goulash, or dumplings. While the portions were a fraction of a normal Western meal, I felt lucky to have something to eat when millions of North Koreans are reportedly going hungry. Malnutrition has led to stunting and mental retardation. Studies done on escapees from North Korea have shown that, on average, teenage boys in North Korea are five inches shorter and weigh twenty-five pounds less than their South Korean counterparts. What I was being served was probably elaborate compared with normal standards for North Koreans, particularly those living outside the capital.

I ate in the guards' area at a wooden coffee table. During the first few days of my confinement, I ate in silence. But slowly I began to feel more comfortable with my guards, and I was desperate for some human interaction beyond the investigation. I wanted to talk to them, and after several days, they seemed to loosen up. During one dinner, I tried to strike up a conversation with Min-Jin.

"You're very pretty," I said. Her cheeks turned a rosy hue.

"No," she replied. "I'm too short."

We started making small talk and chitchatting about everyday things. I stayed away from politics or anything that might be viewed as subversive. Like most twenty-something women anywhere in the world, she seemed most interested in talking about guys and relationships. She asked if it was true that Western men and women like to get drunk at bars and go home and sleep with one another.

"Um, yes, that does happen," I responded with an amused grin. "Where did you hear this?"

"A foreign tourist told me," she replied with a smile.

"Does that ever happen here?" I asked.

"Of course not," she said wide-eyed. "We are not like you in the West."

She then proceeded to mimic a man in a bar by licking her tongue over her lips with a perverted grin. I returned the expression with my own salacious imitation. We both burst into giggles.

She wanted to know what my husband looked like, how long we'd been married, and how we met. It was hard for me to talk about Iain without getting emotional. He was the love of my life. In the twelve years we'd known each other, we'd argued maybe five times. We still were excited to see each other every day. I thought about the last meal we had together. We'd just bought a house so we were trying to save money and eat at home more often. Iain normally did the cooking because, frankly, I'm just not good at it. To surprise him, I had struggled to put together a dinner of beef vegetable soup. I spent the whole day laboring over it, only to miscalculate the amount of salt needed. When we sat down that night to eat dinner, I watched nervously as Iain took the first spoonful. He tried to hide an obvious gag and ended up choking from the excessive salt. He didn't want to make me feel bad and told me it was delicious. When I took a taste, I had to spit it out because it was horrible. It was typical of Iain not to hurt my feelings because he knew how hard I had worked. I insisted that we toss out the soup and head over to Henry's Tacos instead.

I told Min-Jin that Iain and I met at a concert while in college. I couldn't figure out how to describe the sounds of the Chemical Brothers. Just then, Min-Jin started singing some Western songs, beginning with "My Heart Will Go On," by Celine Dion. It seemed that everyone, no matter how isolated their society is, knows the movie *Titanic* and the song that goes with it.

"Do you know what hip-hop is?" I asked. She looked confused. "It's like rap music," I continued.

"Oh, yes!" she replied and jumped up from the couch where she was sprawled out. "Is this rap music?" she asked and began to bounce up and down with her arms spread out. "Yo, yo, yo!" she chanted before keeling over laughing.

We were two young women from opposite worlds sharing a moment of levity. It was the first time I had felt anything other than fear and sadness during my captivity. But then Kyung-Hee walked into the room. Her cold, bemused expression extinguished any bit of cheerfulness in the air. I proceeded to pick at the remainder of my meal in silence as Min-Jin slumped back into her chair.

The next day Min-Jin was noticeably more reserved. When I tried to talk to her, she was curt and unresponsive. I figured she must have overstepped her bounds with me and had either been reprimanded by Kyung-Hee or didn't want to cross the line again for fear of being punished.

EACH DAY I DREADED the daily questioning from Mr. Yee. I would sit on my bed in nervous anticipation of his visits. My stomach began to churn anytime I heard footsteps approaching my room, and I prepared myself to be grilled. With his tape recorder and red notebook in hand, he wanted to know every detail about the story we were covering along the border, including whom I interviewed, what questions I asked, and what was said. Even though it was grueling, I found myself thinking of trivial details that could prolong the process and give my family and government more time to act.

One day, about a week into the investigation, Mr. Yee asked me the question I feared the most. "So you have been to North Korea before. What about your sister? Has she ever been here?"

L AURA'S CAPTIVITY IN NORTH KOREA was hugely alarming for me for another reason. She was being held inside a country that considered me an enemy. In 2003, after three and a half years as a cohost for the daily talk show *The View*, I left the show and started working for an organization that was committed to sending me all over the world to cover stories: National Geographic Television. It was a dream job. In my first few years at National Geographic, I traveled everywhere from Nepal to China, Colombia to Egypt. But there was one country I wanted to visit the most: the place considered the most isolated in the world. Little had ever been reported from inside North Korea. And in my wildest dreams, I never thought I would get a chance to actually go there.

In June 2007, a friend from Nepal, Dr. Sanduk Ruit, a world-renowned cataract surgeon, was invited by the North Korean government to set up medical camps in three North Korean cities. Knowing that I had always wanted to go there, Dr. Ruit asked me if I'd like to be part of his team. The catch was that I was never to say I was a journalist. Ruit convinced the North Korean embassy in Nepal that I was a vital part of his group and that he needed me to document the surgical procedures for training purposes on video. I was therefore permitted to bring a camera and the necessary equipment. I was told I would be the only American inside North Korea at the time, as all foreign nongovernmental organizations and food groups had recently been expelled from the country.

Upon arrival in Pyongyang, we all had to turn our cell phones over to the authorities for storage during our stay in the country. I was told that because North Korea considers itself still at war with the United States and South Korea, cell-phone activity could be detected by satellite. Therefore, no one in the country was permitted to have one. That's what we were told anyway. We had between six and eight government officials assigned to us to monitor our every move—they even stayed in the guesthouses where we were lodged. It didn't take long before I fully understood how the North Korean government maintained its stranglehold over its people. I felt as if everyone was watching me, and it seemed everyone was watching one another.

For the twelve days we were in North Korea, we worked in hospitals in the cities of Pyongyang, Pyongsong, and Sariwon. Though we were in operating rooms for most of our visit, we were given a unique window into North Korea's health-care system and medical facilities. I would have to describe the latter as antiquated and basic. Dr. Ruit had to bring all of the equipment he needed, including generators to deal with the frequent power outages that occurred throughout our visit.

Cataracts develop for many reasons, but the most common are those induced by old age, malnourishment, and excessive exposure to ultraviolet light. Although the camps were set up to treat cataracts, the number of people who showed up with other maladies like head pain, body aches, and dizziness, among other things, was an indication of the extreme conditions and lack of proper treatment.

In the developed world, cataract operations are commonplace—people typically have cataracts removed in the very early stages before they start to impede vision. In poor countries where medical treatment and technology are not readily available, people with severe cataracts can go totally blind. When this happens, they cease being productive members of society and become a strain on their families and communities. Often that leads them to be cast out and ostracized.

While we were in the country, thousands of people came to have their eyes checked, including young children. Dr. Ruit or one of his trainees performed the twenty-five-minute operation on those with bona fide cataracts. In the end, a thousand surgeries were performed over twelve days. People who had been blind for as long as ten years had their sight entirely restored. It was miraculous. But it wasn't Dr. Ruit and his team who were thanked for performing the work necessary to give people their vision back.

Once their operation was complete, patients were asked to go into the large waiting room to rest for at least twelve hours. Two hundred people would be in the room waiting for their eye patches to be removed. One by one, Dr. Ruit's staff peeled away the bandages from the formerly blind patients' eyes. Once the bandages were removed, every single person would become hysterical and rush to the gigantic portrait of North Korea's despotic leader, Kim Jong Il, to thank him for curing them of blindness. Like robots, hundreds of people collapsed in tears before the image of Chairman Kim, whom the North Korean people affectionately call "Dear Leader." It was as if the doctor weren't even there.

I was purportedly documenting the teaching process, and I asked the North Korean officials if I could follow a recovered patient home to see how she navigated with her sight restored. Initially, my inquiry was met with looks of confusion. Everyone seemed to be wondering, *Why would she want to do that?* But to my surprise, after a lengthy discussion, they agreed.

I was taken to the home of an older woman who lived with her son, his wife, and their two young daughters. It was a decent-sized one-bedroom apartment in a three-story walk-up. There were no beds in the home because everyone in the family slept on the heated floor in the living room. I was told that this was typical. They had a television and a large stereo system that looked like a boom box with big speakers. There wasn't a family photo to be seen, only those of the Dear Leader and his father, the "Great Leader," Kim

Il Sung. All over the apartment, there were small photos and large photos of them, some translated into watercolor paintings and others into oils. There was the Dear One riding horses, leading a parade, and walking among flowers. I asked if they thought the Dear Leader ever did anything wrong. The entire family seemed stupefied, and my North Korean government escorts looked completely befuddled. It was obvious that my question had never been asked before, so they just sat and stared at me for five of the most uncomfortable minutes of my life.

Sariwon was the last stop on our three-city medical mission. Several hours south of Pyongyang by car, it was a nice, well-maintained city with clean, wide streets, parks, and large sprawling fields. In the first two cities we'd been in, I was told that I could only jog laps around the inside perimeter of the guesthouse property. Sariwon was less populated, so the government officials said I could jog on the road around the hotel as long as I didn't veer off onto other streets.

It was about 7:00 A.M., and the air was brisk. I could smell smoke from the fires that people burned to heat their homes at night. At this early hour, people were walking to school and to work. Even though the streets were wide and vast, the whole time I was there I saw few bicycles and even fewer cars. I jogged past three girls who were clearly fascinated by me. They wore monochromatic tracksuits and their haircuts were short and boyish. They looked like athletes.

On my third lap around the hotel, the girls started following me as I ran. All three kept up with me for about ten minutes until two of them dropped off. The tallest of the three girls maintained a decent pace, even passing me a few times. I started running faster, and then she picked up the pace and passed me. Whenever I'd step ahead, the girl would speed up. And as soon as I passed her, she'd run past me. At a certain point, we were running at full speed. We were going so fast that people in the town rushed to watch us, and then it became a race. The whole time I was thinking, *I'm running for America!* I could see in her face that she was running for her country. Neither of us would give in—except at a certain point, after completing an entire lap at full speed, we both stopped simultaneously. We were panting so ferociously we could hardly stand up straight. Between breaths, we caught each other's eyes, and all of a sudden the two of us burst into laughter. For a moment we were just two young women having a funny moment. It was nice.

On the night before leaving the country, I was asked to hand over all the tapes I had shot of the "training," so they could be individually screened by government officers. I had anticipated this, so I hid a couple of tapes in the underwear pouch of my suitcase. The resulting documentary revealed an unprecedented and critical look into North Korea's crippled health-care system and the extreme indoctrination under which the people there live.

After breakfast on the morning of my departure, I got into a conversation with one of the North Korean official minders assigned to watch over us. Of all of them, he was the sternest and most defiant about North Korea's strength in the face of hardship. He was nearly six feet tall, with sharp cheekbones and a military-style crew cut. He was called Kwon.

"We are a small country, but we are all we need," he explained.

He told me that as soon as we left the country, he and the rest of the minders would join their brothers and sisters and head to the countryside to harvest rice. Every citizen of North Korea, whether a government official, bus driver, or street cleaner, is obligated to work in the fields during harvest season. Kwon said it was "for the good of our nation."

"In your country, all you think of is yourself," Kwon continued, then launching into a tirade about the sorry state of affairs in America.

"Your president [it was George W. Bush at the time] invades

other countries for their oil so Americans can drive their big fancy cars, and then he calls other countries 'evil.'" Kwon was referring to President Bush's characterization of North Korea as part of an "Axis of Evil" with Iran and Iraq in his 2002 State of the Union Address. Kwon went on to ask, "Who is the evil one?"

I think I surprised him when I told him that, though I was an American, I was very much opposed to the war in Iraq. But at least, I said, "I can express that publicly in my country."

After a slight pause, Kwon retorted with something that surprised me. "My country and your country had a chance for more normalized relations under Clinton. We like him."

"Who likes him?" I quizzed.

"The North Korean people," he said.

I noticed that he, like so many of the North Koreans I met, never talked about his individual likes and dislikes. It was always about the collective; it was about their country.

Toward the end of his second term, in October 2000, then President Clinton sent Secretary of State Madeleine Albright to Pyongyang for a meeting with North Korea's leader, Kim Jong Il. She was the first-ever American secretary of state and the highest-level U.S. government official to visit the Communist country. The trip was meant to persuade Kim to stop developing, testing, and exporting missiles. Albright was also setting the stage for what would have been an unprecedented visit by President Clinton to North Korea if a deal could be hatched. It was a high point in U.S.–North Korean relations. But Clinton's term came to an end before he could make the trip, and his successor, George W. Bush, had a very different view of how the United States should deal with North Korea.

As our bags went through a final check at the airport, my heart was racing. Clear. When the plane lifted off from Pyongyang, I breathed in deeply and slowly exhaled. I was out. I never expected to have any further dealings with the government of North Korea.

Given how critical my 2007 documentary was of the North Korean regime, I now thought I should try to reduce the possibility of a paper trail to me right away. Even though the North Koreans had probably already seen my documentary, I didn't want to make an already contentious situation worse for Laura. One of my early calls was to the president of National Geographic, Tim Kelly. Tim hired me to host the *Explorer* series in 2003 and had become a good friend. I told him I was very concerned that my work in North Korea would adversely affect my sister's situation and asked him to pull my documentary off its programming schedule and stop selling copies of the video. Within the day, National Geographic had imposed a total moratorium on the sale and airing of my *Inside North Korea* film and had it and all related clips pulled off of YouTube.

I also put out calls to my contacts in the media, which had grown quite extensive over my many years working in broadcast television. I called the presidents of a couple of cable news networks as well as a number of producers and correspondents. I acknowledged the need to report the news, but asked if news directors could limit the coverage of my sister's detainment, as we were dealing with an extremely unpredictable actor in the North Korean government and we were deeply concerned about antagonizing him. In the early days of Laura and Euna's detainment, we didn't want to give the North Koreans reason to think they had a political bargaining tool in the girls. Every news organization has dealt with emergencies associated with sending correspondents into the field; therefore, my colleagues in the industry were very supportive of my request, and aside from reporting the basic facts about the detainment, almost all the major TV news outlets reported our story quite minimally.

# the visit

∽∾ LAURA

WHEN MR. YEE ASKED me about whether or not Lisa had been to North Korea, I knew I couldn't lie. "Yes," I replied nervously. "She came as part of a medical delegation working on a documentary." While I had been nervous about the North Koreans learning about my past project there, I was even more afraid of them finding out about Lisa's work. The reports I helped produce in 2002 were benign compared with Lisa's documentary.

"Why didn't you tell me this earlier?" he asked sternly. "Did you think we wouldn't find out? We know everything."

He then pulled out a dossier he had on Lisa's visit there.

Lisa's documentary for National Geographic was highly critical of Kim Jong Il's regime, and now the North Koreans saw her as an enemy of their country. Mr. Yee accused Lisa and me of having hostile intentions toward his country. He told me that Euna had a chance of going home because she was cooperating fully, but I, on the other hand, might have to spend the rest of my life in a North Korean prison.

I was so scared I could barely breathe. If Mr. Yee was trying to terrify me, he was succeeding. How had things gotten to this point? With each moment, my life seemed to be slipping further and further away. It was completely mind-boggling. Yet considering how paranoid the North Korean government is, I could see how they might think two sisters who were journalists could be out to topple the regime. This was made even worse because Lisa had worked on a scathing report about the North Korean system, and now I was being investigated and potentially accused of espionage in the DPRK. I tried to convince Mr. Yee that our assignments in North Korea were unrelated and not part of some intricate plot.

"Tell me about her documentary," he said. "What did she film?"

"I don't recall exactly," I said, lying. "All I remember is that a surgeon came here to remove people's cataracts. My sister had worked with this doctor before in Nepal and wanted to help him raise money for the work he does to restore people's vision. She only wanted to help him."

"Are you and your sister trying to overthrow the North Korean government?" Mr. Yee angrily asked.

"Absolutely not," I replied. "The fact that I'm here right now is entirely coincidental."

That night, I lay in bed and thought about how Lisa might be coping. I knew she must be terrified and wondering if North Korean authorities had linked her to me and, if they had, what that might mean for me here. I thought back to the morning when she was preparing to go to North Korea and how concerned I was that officials might discover she was on television. I was shocked that her visa to travel with the medical team was approved at all, and I warned her to be ultracautious. We joked that she better not be kidnapped by Kim Jong Il and be taken for one of his wives. Thinking back to that conversation gave me chills.

During that trip, she managed to call me once from Pyongyang. I knew her phone was being bugged, so I was careful about the questions I asked. I was just happy to hear from her.

"Are you having a good time?" I asked.

"Yes!" she replied excitedly. I knew instantly she was putting on an act. "It's really interesting here. Anyway, I can't talk long. I just want you to know I'm okay, and I'll see you soon."

I remembered hanging up after that call and thinking that if I lost my sister, it would be like losing a part of myself. I wasn't able to put my worrying aside until Lisa landed safely in Beijing a few days later. Now I wondered if I might be punished not only for my actions, but for hers a few years earlier.

THOUGH LISA AND I were brought up by a devoutly Christian grandmother, I've never been particularly religious. But like most people in times of crisis, I looked to a higher being for guidance and assurance. I began to talk to God each morning, asking him to help me get through another day. I prayed to the Lord to give me and Euna strength to endure, to watch over our families, and to give my interrogator, Mr. Yee, compassion. I found solace in this simple ritual, and it made me feel less alone.

For several days, I'd been wondering where I was being held. I figured I had to be somewhere in or near Pyongyang, but I wasn't certain. The thick mustard-colored curtains in my room had remained closed since the day I arrived, and I wasn't allowed anywhere near the window. "Lord, show me a sign," I whispered out loud. "Send me a signal that things are going to be okay." I felt slightly silly for making such a clichéd request of God, but within minutes, the entire sheet of drapes along with a long metal rod fell to the floor with a thunderous crash. I popped up, careful to stay in bed so my guards wouldn't think I'd had anything to do with the falling curtains. For the first time in days, I could see outside. I marveled at the birds and the sky.

Trees surrounded most of the compound, but through the window, I was able to make out the top of the famous pyramid-shaped Ryugyong Hotel, which the builders had intended to make so high it would be the world's tallest hotel. I had seen the building when I was in Pyongyang in 2002. Its 105 stories and 1,100 feet of glass make it an impressive sight. The hotel was started in the late eighties, but structural problems and a lack of resources and money caused it to be left unfinished and empty. Back then, our North Korean guides said the building was still under construction, but the absence of workers and cranes made it clear the project had been abandoned. Rather than becoming a source of pride for the country, the structure has become an embarrassment and a symbol of North Korea's faltering infrastructure.

This is how I knew I was in Pyongyang, and not too far from the city center. I had never in my life challenged God to give me a sign. Now the first time I asked for one, he seemed to have answered. The guards rushed in to fix the drapes. I wanted to help them and rose from the bed so I could lift the fabric from the ground. They were too flustered to order me away and they seemed to appreciate my help. All the while, I stole glances outside, taking in as much of the scenery as I could. There wasn't much to look at except for a row of tall trees and some shrubs that were obstructing my view of a building down below the compound. But for me, seeing this small glimpse of nature was a little gift from God.

About a week into my detention in Pyongyang, which was close to two weeks since my capture, the guards asked me if I wanted to take a warm bath. The bathroom had a tub, but it was always being used to store water for those frequent times each day when the water shut off. Sometimes these outages lasted several days. When that happened, the water in the tub could be used for basic needs such as flushing the toilet and brushing teeth. In any case, there was no hot running water, so the tap water was too cold to use for washing. For the past week, cleaning myself had involved soaking a towel in the frigid water, bracing for the forbidding cold, and rapidly scrubbing myself. I welcomed the opportunity to take a warm bath.

The way they heated the water was to place an electrified metal rod into the filled tub. I was told it could take several hours to heat, but after five hours the water was just barely lukewarm. It had a number of hours to go before it was bearable, but then the electricity shut off. The guards looked defeated. I thanked them for their efforts and told them I didn't need to take a bath. Truthfully, the last thing on my mind was my hygiene. As time went on, I developed a system by which the guards would allow me to heat water in a little electric kettle. Just getting this small amount of water to boil took nearly thirty minutes. I would mix the hot water with some cold water from the bathtub, which provided just enough warm water to splash onto my body and rinse off.

When I wasn't being interrogated, I was usually curled up in bed. I had nothing to do but think. I also listened intently, straining to hear the slightest sound that might indicate that Euna was also being held nearby in the compound. I faked loud coughs or sneezes, hoping she might hear me. Once in a while, I would ask Min-Jin about Euna.

"How's my friend doing today?" I asked nonchalantly, as if I already knew she was somewhere close.

"Who's your friend?" she replied with a blank look.

"Her name's Euna," I said dejectedly. "She's Korean. I think you'd really like her."

I couldn't tell if she really knew who Euna was or not. If Euna wasn't being held in the same facility, where was she?

To keep warm and to occupy my mind, I walked in circles around my room, sometimes hundreds of them a day. I thought a lot about my parents. My heart would break every time I pondered how worried they must be about me. Aside from their shared love of shellfish, our parents have nothing in common except Lisa and me—we are everything to them. I wondered what they were doing and how they were dealing with all of this. And were they together?

Our parents' relationship had mellowed a great deal since the time when Lisa and I were kids. Lisa teases by saying we're closer as a family now than we were when our parents were married. The four of us—Mom, Dad, Lisa, and I—have even gone on a couple of family vacations together over the last few years. We laugh when our parents joke about which one of them used to snore the loudest. When we were kids, however, the thought of our parents ever becoming friends was simply unthinkable because all I can remember about their life together is the fighting.

When I was four and Lisa was seven, my father was driving all of us down a slick road in the dark of night. I was in the front seat of our tan Buick Oldsmobile between my mother and father. Lisa was sleeping across the backseat. I nestled my head in my mother's lap, but I wasn't trying to fall asleep. I wanted to drown out the yelling Lisa and I had become so used to hearing. But this evening's conversation was different. Our mom was telling our dad that she wanted to move to Los Angeles and live with her sister there. She wanted a divorce.

The screeching sound the tires made on the pavement seemed to go on forever. I struggled for breath as my father's weight enveloped me. He had taken his hands off the wheel and was reaching for my mother. I tried to scream, but nothing came out. Suddenly I heard Lisa's voice from the back of the car, shouting, "Daddy, stop! Please stop!" Her little hands reached over the seat, grabbing my father's hair with all her strength. Lisa's vigor worked, and Dad calmed down. Then he just started crying. We were all crying.

Months later, Mom moved to Los Angeles. My dad's mother was living with us at the time, but Lisa became my protector and I hers . . . even when it didn't exactly work to my advantage.

One day Lisa—age eight—was in the bathroom for a long time. I kept coming back and would stand outside the door waiting for her to come out. She didn't. Then, from my room, I heard a loud crashing sound from inside the bathroom. Running to the door, I screamed, "What happened, Li? Are you okay?"

Lisa opened the bathroom door and pulled me inside. She had spilled foot powder all over the room; whiteness blanketed the sink and the dark blue carpet. I heard Dad's footsteps coming up the stairs.

"What the hell is going on in there? Open the door!"

We opened the door, and he looked around to see everything covered in powder.

"What happened here?" he asked angrily. "Who did this?"

Silence. Then Lisa looked over at me and in her little voice said, "She did," and pointed right at me.

What? I couldn't believe she was blaming me. I knew she just didn't want to get into trouble, but how could she? Then my dad looked down at me and asked, "Is it true, Laura? Did you do this?"

I looked up into his eyes and then looked over to my sister, who was looking down. "Yes . . . I did."

Dad immediately sent me to my room and made me sit in the dark. An hour later Lisa came in and turned on the light. She looked sullen and forlorn as if she'd been crying. I could tell she felt guilty about telling Dad that I spilled the foot powder.

"Thanks for taking the blame for me, Lau," she said. "Why didn't you stand up for yourself?"

I didn't say anything.

Lisa then came over and put her arms around me. "I promise, Baby Girl, that I'll stand up for you as long as I live."

THE UNITED STATES HAS a diplomatic relationship with almost every country in the world—some exceptions are Iran, Cuba, and of course North Korea. Without some form of diplomatic relations, the leaders from each country cannot just pick up the phone to call one another. It seems entirely illogical and even archaic, but that's just the way it is. If the United States ever needs to send a message to North Korea—or vice versa—it has another country, Sweden, as the official liaison. The Swedish ambassador to North Korea is named Mats Foyer.

Linda from the State Department suggested that we write a short letter to Laura, and she would send it to her through the Swedish ambassador. She said she would e-mail the note to Ambassador Foyer, and he would print it out and deliver it to the North Korean Foreign Ministry in Pyongyang. It was worth a try. My parents, Iain, Paul, and I huddled in my mom's den for an hour and crafted a meticulously worded letter from all of us.

It was helpful to have Paul's insights because his father, Won Ryul Song, was from Pyongyang and had fled south to Seoul from the north in 1946 when Communism was starting to take root there. It was seven years before the two Koreas were divided. As my father-in-law told it, it was a time when young men were being rounded up and forced to become part of the Communist movement or face being sent to a labor camp or, even worse, to death. He has not been back since. As I learned from him and from actually being there myself, there is perhaps no culture on earth more obsessed with respect and the idea of saving face than North Korea's. We had to be deferential and even apologetic in our letter because we knew that my sister's eyes would not be the only ones to read the letter. We were certain that it would be seen and scrutinized by those holding Laura. No matter what actually happened on the border of China and North Korea, it was imperative that we express great remorse and our apologies. Anything perceived by Laura's captors as hostile or accusatory could make matters much worse for her.

One thing that we made sure not to do was refer to God or prayer in our letter. There is one group of people for whom the North Korean government has more contempt than it has for Americans. They are the people the regime believes are fervently trying to overthrow it: Christians. Christian groups in South Korea and along the Chinese border lead the charge in protesting North Korea's abysmal human rights record. It was, in fact, a South Korean Christian pastor who had helped Laura's team arrange interviews and plan the shooting schedule. There are reputed to be thousands of underground Christians working along the border trying to help people escape from North Korea. We were therefore careful about not seeming to be aligned with any such groups. We suggested that Michael also write a letter to Euna using the same precautions and guidelines.

Ambassador Foyer was our only hope of getting information to Laura and Euna. After we e-mailed our letter to Linda, she forwarded it to the ambassador. We had to wait to see if it would actually get to Laura.

The other way the United States and North Korea can communicate is through the latter's Permanent Mission to the United Nations based in New York City. The U.S. State Department occasionally sends messages through a North Korean diplomat based in New York named Minister Kim Myong-Gil. In diplomatic circles, Minister Kim is known as the "New York channel." Right after we learned of Laura's capture, my mother found Minister Kim's information online and sent a letter, fax, and e-mail to him every day thereafter. She also left daily phone messages. Mom would call his voice mail and make lengthy sobbing declarations of the extensive pain the North Korean government was inflicting on her by holding her child prisoner. From another room I would hear the anguish in my mother's voice when she'd call. There were a few times when she wouldn't even be able to speak, she would just cry into the receiver and then hang up.

## ∿ LAURA

ONE EVENING, TWO WEEKS into my detainment, Mr. Yee brought me a box wrapped in decorative paper from one of Pyongyang's hotels. "These are toiletries that are being provided to you by the DPRK and by me because I am in charge of you," he said. "They are not being sent to you by your government. Do you understand?"

"Yes," I replied. "Thank you so much."

The box contained a facial lotion set with three different types of creams and a toner made from ginseng root. I could sense that the interrogator's comment was intended to let me know that his government was benevolently caring for me and that I should be grateful for this treatment.

"It looks very nice," I said appreciatively. "Thank you for being so kind."

He then pulled out a manila envelope. The top left corner had some sort of official seal with English letters. My heart began to thump wildly, and I strained to make out the words on the seal. I hoped this would be a message from the U.S. government or my family.

"This is an envelope from the Swedish embassy," Mr. Yee said. "Because the United States and North Korea are still at war, you have no official representation here. So the Swedish government acts as an intermediary body for the two countries. Do you understand?"

"Yes, sir," I replied, growing increasingly anxious to see what was in the envelope.

"Here are some letters from your family and colleagues," he said, handing over the thin packet.

I rushed into my room, sat down on the bed, and tore open the envelope. The letter was an e-mail dated Saturday, March 21, which was the day I was being transferred to Pyongyang. Only a week had passed, but it felt like an eternity. I burst into tears just looking at the typed words on the paper.

> *Our Dearest Laura,*
>
> *We miss you so much. Please be strong and know that everyone is doing all they can for you and Euna. We are holding you in our hearts every second of the day and night. Friends and family from every corner of the world are sending positive thoughts. We have been moved by how many people have expressed their love for you; your Facebook page is overflowing with messages of support . . . We know that those in whose care you are, are not harming you and that you are safe. Please be respectful to them. Stay strong, Baby Girl. We know we'll see you very soon. We love you so much.*
>
> *Love, your family, Dad, Mom, Iain, Li, and Paul*

I knew my sister had written this letter on behalf of my family. "Baby Girl" was what she often called me. I also noticed that Lisa had not used her full name but had signed the letter "Li," probably because she hoped it wouldn't catch the attention of the North Korean authorities. I read the words "We know we'll see you very soon" over and over. By the next morning I had memorized the letter completely. Its optimistic tone gave me some hope that things might work out. At the same time, the message was vague and said nothing specific about what was being done to bring us home. Were these just veiled words to keep my spirits up? How could my family be so reassuring that we would be reunited soon when the letter was sent just days after I was captured? Already a week had gone by, and Mr. Yee was giving me the vibe that going home "soon" was far from likely.

The next day, he came into my room and told me to wash my face and make myself presentable because I would be meeting with someone. Astonished, I moved to the edge of my seat, wanting to hear more. I was going to be taken to see Mats Foyer, the Swedish ambassador to North Korea. Mr. Yee explained that my time with the ambassador would be extremely limited, so I should use the time wisely.

"He will want to know how you've been treated," Mr. Yee said. "We have not harmed you, have we?"

"No, no," I replied. "I've been treated fairly."

He nodded in approval. "Because of your crimes, your situation is very serious. This is the time to tell him that your government must do something."

My body was trembling. Was there news? Would he be bringing additional messages from my family or my government? I was filled with excitement but cautious about letting it overtake me. I had so much I wanted to convey, but how could I do it with North Korean officials in the room? I particularly wanted him to know about the violence, but given my conversation with Mr. Yee about how I'd been treated, I knew I couldn't just tell the ambassador outright about my condition. My hands were shaking. When my guards were not paying attention, I carefully slid the bandana on my head up just enough to reveal a small strip of the bandage that was covering the large gash. I had to find a way for the ambassador to notice the gauze strip so he would tell my government that action was needed urgently.

A driver arrived at the compound and I was directed into the backseat of the car, accompanied by Mr. Yee and Mr. Baek. They told me we were going to the Yanggakdo Hotel, where I would meet the ambassador. I'd been conditioned to hold my head down while I was being transported between locations, but this time I asked Mr. Yee if I could look out the window.

"Yes, go ahead," he answered.

I peered out the window as we made our way to the hotel. Pyongyang is the country's pride, and living there is a huge privilege, one that's reserved for the most elite, loyal citizens. Mr. Yee explained to me that after the city was demolished during the Korean War and riddled with land mines by the enemy forces, the area was deemed too dangerous to build on. He took great satisfaction in telling me that despite these setbacks and being isolated by sanctions, North Korea had erected an impressive city of skyscrapers, parks, stadiums, and monuments out of the rubble. I did find Pyongyang to be an attractive city, but it's clear that they've not been able to invest much in upkeep and maintenance. There don't seem to be many buildings that have gone up in the last decade, and up close, the existing ones look to be in disrepair. Many of the sidewalks are in ruins and unusable.

I noticed that there were more cars on the road than during my previous visit to Pyongyang in 2002, when the streets were empty of vehicles. Now, at an intersection, I'd see two or three other cars lined up, which in North Korea constitutes a traffic jam. At the center of each intersection was a pretty woman dressed in a blue uniform made up of a skirt, a coat, and a large military-style cap. Each held a flashlight-type wand, which she used to direct the cars. These human traffic guides were necessary because the lights are unreliable due to the frequent power outages. Not only are cars too expensive for the average citizen to own, but it is difficult to be granted a vehicle permit. Most people walk or take public transportation. We passed the local train station, where groups of men and women in army uniforms were squatting here and there waiting for the train. I noticed that there were more shops and restaurants than before, and small beverage kiosks had been set up on the sidewalks. Unlike most capital cities across the globe, Pyongyang is devoid of commercial advertisements. Any signage is devoted to Communist propaganda sayings or giant portraits of Kim Il Sung and Kim Jong Il. Everything was extremely orderly. People weren't lingering or conversing with one another on the streets. They just seemed to be going about their business.

The towering Yanggakdo Hotel, one of Pyongyang's finest, sits on a small island within the city. We came into the vast lobby, which was empty of any tourists or businesspeople, and I was brought to a small conference room for the meeting. The ambassador was tall and lanky, with a kind and gentle demeanor. When he embraced me, I was overwhelmed with emotion. For the past two weeks, I'd felt like I was living in a parallel universe, with no way of connecting to the life I once knew. But through Ambassador Foyer's eyes, my family would in turn be seeing me. Tears trickled down my cheeks as I struggled to compose myself. I had been told I would have only ten minutes with the ambassador, and I wanted to make every second count.

The ambassador explained that his questions had to be of a consular nature, such as how was my health and how was I being treated. I subtly directed my eyes upward to signal the bandaged wound I received from the butt of the rifle during the time of my arrest. But with several North Korean officers there, monitoring every word, I found myself saying, "I'm being treated fine."

I told the ambassador I had apologized for my crime of trespassing into North Korean territory, but that my situation was very grave because of the documentary report we were working on and because my sister had worked on a film in North Korea. He and a colleague from the Swedish Embassy jotted down notes as I explained my situation.

I knew that if I said anything that angered the officials in the room, it would be bad for me. But there was one thing I wanted to communicate. "We did cross the border very briefly," I explained. "I'm very sorry for that. I've expressed deep remorse for my actions. It all just happened so fast. We were running back to the China side when we were arrested on Chinese soil."

I knew it was risky telling him we were in Chinese territory when we were taken. I didn't know how the authorities monitoring me would react. But this could be my only chance with the ambassador, and I thought this bit of information might give the U.S. government some leverage in dealing with North Korea. Or perhaps it could help the Chinese government lend a hand to the United States and put some pressure on the North Koreans to release us.

The ambassador looked at me tenderly and told me to be strong. He asked about my health and my ulcer. I assumed my family had made him aware of my condition. I mentioned some stomach pains that I had been experiencing, and he said he had a package of medication along with books, snacks, and toiletries that had been sent via the U.S. Embassy in Beijing. He inquired if there were any items I needed.

"It would be great to get some more letters," I replied.

He explained that by international agreement I had a right to send and receive letters, and said that he would ask to see me regularly.

He assured me that my employers, including Vice President Gore, were working on the case. I asked if Secretary of State Clinton was aware of the situation. He affirmed that she was not only concerned but also involved.

One of the North Korean monitors signaled that our time was up. I thanked the ambassador and rose to give him a hug. I didn't want to let go of his thin bony frame. He was my lifeline, my only link to the outside world.


## ⟨⟩ LISA

L INDA McFADYEN, OUR FAMILY'S State Department contact, told us that the very moment Ambassador Foyer learned of Laura and Euna's detainment, he'd begun trying to see them. He even delayed a number of overseas trips he was supposed to make so he would be available if given the chance to see the girls. The ambassador knew he was my sister and Euna's only connection to their government and their families, and he took this role very seriously. Linda said he had been calling the North Korean Foreign Ministry every day and often showed up at their offices demanding to see the girls.

On March 30, Linda told us that Ambassador Foyer had been granted a visit with Laura and Euna and that he'd been allowed ten minutes with each of them. This was a huge development because this was the first time we'd heard anything about their condition since the arrest. We'd already endured two torturous weeks of nothing but silence from North Korea, during which we ran through every worst-case possibility of how they might have been treated. Ambassador Foyer was the first and only non–North Korean to lay eyes on my sister.

He told the State Department, and Linda told us, that Laura was physically okay but very, very scared. He went on to say that she had been extremely talkative, as though she was trying to utilize productively every second she had with him. She told the ambassador that, yes, they had "touched" North Korean soil but were back across the border and in China when they were taken by the North Korean guards, who then took her back across the frozen river. Laura had confirmed what we already knew from Mitch. Though I was still hopeful that Laura and Euna's apprehension on Chinese soil might give us some leverage if China would agree to pressure North Korea, I remembered what Governor Richardson had advised—that getting China involved would only infuriate the North Koreans. Not only that, we had no idea whether China would cooperate. Our family decided it would be best to keep this information under wraps, at least for now.

Lastly, Linda said something that made my heart sink—she said the North Koreans were displeased with the documentary I made for National Geographic.

This was all so much to process. Were they holding what I did against her? They had figured out that Laura and I were sisters, and I could only imagine what they were hypothesizing about us. I felt like I was having a breakdown, but I knew I had to hold things together. Whatever I was going through paled in comparison with what Laura was enduring. I needed to use every ounce of energy within me to concentrate on getting her home.

### ✌ LAURA

THE NEXT DAY, MR. YEE asked me to recall what I had said during my meeting with the ambassador. I tried to go over the conversation, emphasizing that I had told the ambassador I was sorry for my actions and was very remorseful. But Mr. Yee just scowled.

"Why did you tell him you were apprehended in China?" he scolded. "What did you think you would accomplish by telling him that? My bosses are very upset with you!"

"I am so sorry," I said, my voice shaking in fear. "I didn't mean to anger anyone. I was just telling the ambassador what had happened, and that is what happened."

Exasperated, he shouted, "You violated DPRK law by coming into our country. We can chase you anywhere to arrest you. That is our right!"

"Please," I begged, "would you please let me see the ambassador again? I will apologize and say anything you want me to say. I don't want to upset your bosses. I only want to be cooperative."

"Do you think you can just see the ambassador anytime you wish?" he continued. "We gave you a courtesy visit. We don't need to let you see him again."

I felt like I'd just been punched in the face. I was afraid I had squandered any future chance of seeing Ambassador Foyer. Nearly two long months would pass before I was allowed to see him again.


GETTING THROUGH EVERY MINUTE of each day was a psychological battle. Power and water outages were frequent, and sleep became a challenge, with frequent nightmares. Some nights I would dream I was back at home but had only twenty-four hours to get the U.S. government to sign a peace treaty with North Korea and remove our troops from the border or I would be sent back to Pyongyang to face my execution. I also had recurring nightmares of Euna being tortured by cruel soldiers. Often I'd wake up in a panicky sweat.

To calm my nerves, I hummed the words to the song "Edelweiss" from *The Sound of Music,* which my mom used to sing to Lisa and me as kids to put us to sleep. *Just stay healthy and hopeful,* I would tell myself, given my ulcer. But as optimistic as I tried to be, I was frustrated by how helpless I felt.

One day, not too long after the meeting with Ambassador Foyer,

Mr. Yee brought two blue satchels. He put them on the desk in front of me and said the ambassador had sent me some items. In the bags were a toothbrush, toothpaste, feminine products, and snacks such as potato chips, bread, peanut butter, chocolates, and a six-pack of Coca-Cola. Mr. Yee then brought out some medication that had been sent by the U.S. Embassy from the Beijing Hospital, including medicine for my ulcer. I'd already experienced several ulcer flare-ups. The area just above my abdomen would burn in pain. I was grateful to have something to alleviate these throbbing episodes. But what I wanted most were the books. The interrogator pulled out a collection of classic novels including Mark Twain's *The Adventures of Huckleberry Finn,* Faulkner's *The Sound and the Fury,* Hemingway's *The Snows of Kilimanjaro,* the entire works of Jane Austen, and Ian McEwan's *Atonement.* There was a letter from Ambassador Foyer along with the books, explaining that the U.S. Embassy in Beijing had selected the novels. The constant, day-after-day interrogations and isolation had left me desperate for some sort of escape. But to my dismay, Mr. Yee returned the books to the bag and pushed it aside.

"Can I please have a book, just one?" I pleaded.

"You are under investigation," he retorted. "You need to be thinking about your crime, not reading. Also, we cannot guarantee the safety of all of these items, especially the food. While I'm sure the ambassador has good intentions, these things have changed hands many times. There are people in your government who may want you dead so the U.S. can start a war with us. We cannot trust that these items are safe." He put the food back in the bag, leaving me with the basic toiletries and medication.

A T THE END of the first letter we wrote to Laura, all my family members signed their full names. I just wrote "Li." While this is what Laura always called me, I didn't want to use my full name because I feared that direct association with me might be dangerous for her. Even in the weekly packages Iain would later send to her, he never included any photographs of me. We didn't want to give my sister's captors any kind of reminder of what I had done there.

Ambassador Foyer was told that our letter had been delivered to Laura, and he communicated to Linda that if we wanted to continue sending letters through him, he would drop them off right away to offices of the Foreign Ministry. He told her he was surprised that the North Koreans said they would pass on the letters to the girls. Linda is convinced that our respectful and apologetic tone in the first letter opened the way for future letters to be delivered to Laura.

I started writing to my sister every other day, if not every day. In each letter, I addressed her with the same greeting I'd used for her all our lives—"Dear Baby Girl . . ." Every letter I wrote was crafted with extreme caution. I never lost sight of the fact that the words I wrote would likely be read by multiple people. I missed my sister so much that the letters were painfully hard to write. I would stare at the blank screen for half an hour at times with tears running down my face at the thought of her out there and alone. I wanted her to know that we were doing everything in our power to get her out. I encouraged her to try to still her mind and take care of her health so she wouldn't exacerbate her medical condition. Sometimes I tried to be funny. I told her that every one of my ex-boyfriends had called or e-mailed to check on me and offer support. Even in captivity I knew she'd laugh about one guy in particular whom she loathed and made me break up with.

At times I was so angry I wanted to make threats to those reading my letters. Once I wrote that I was preparing to mobilize thousands of people from all over the world, using Facebook, to walk across the Chinese–North Korean border if my sister weren't let go. My husband promptly scolded me before I could hit SEND. Thank goodness he stopped me. Paul was always making sure that I acted reasonably and not irrationally. There was just too much at risk. So most of the time I tried to comfort Laura in my letters and take her back to some of the fun times we'd shared together.

IN ONE LETTER, I WROTE:

> Dear Baby Girl,
>
> Every day I think about how you are spending your days: do you go outside, can you watch TV, do you talk to anyone? . . . I was thinking of the days before my wedding when you and I went to get a spray tan. Remember when we stood there in the buck while that woman sprayed every inch of us? There we were in our shower caps, freezing and laughing . . . I love you so much, Baby Girl, and am desperately hoping that the government of NK will show mercy and allow you to come home . . . I need you and I will not stop until you are home with me.
>
> I love you more than anything in the world, Li

As hard as they were to write, these letters were the only way I could communicate with my best friend. They were my way of letting my sister know that I was fighting for her. I never got enough courage to use my full name at the end of any of my letters so I continued to sign off as "Li."

ALTHOUGH I WAS SUBJECTED to hours-long interrogations almost every day, there were stretches, usually over the weekends, when Mr. Yee didn't come to see me. I didn't know for sure, but I suspected Mr. Baek and Mr. Yee were staying at the compound because they would appear at different hours of the day and night. Their absence during the weekends indicated to me that this was when they went home to their families.

During these periods, I'd sit in utter silence. Though Min-Jin had become less approachable after our conversation about music and dating, I tried to reignite some sort of relationship with her and Kyung-Hee. Min-Jin was trying to get better at speaking English, and she often had her head buried in a thick Korean-English dictionary. Kyung-Hee was studying Mandarin Chinese and sometimes read words and sentences out loud. Fortunately, my very elementary knowledge of Mandarin was more advanced than what she was reviewing. I told them I'd be happy to help them with both their English and their Chinese.

This created an opening, because they were both determined to improve their language skills, and they began to look at me as a rare and valuable resource. I was a live tutor, available to them twenty-four hours a day.

Once, while I was walking around my room for exercise, Min-Jin started following me as she practiced some new words she had just learned. "Toothbrush," she said out loud, though it came out sounding more like "toosebrush." Native Korean speakers often have a difficult time pronouncing the *th* sound, which is nonexistent in the Korean language. I mouthed the sound *th* several times, showing her how the tongue touches the front teeth. She repeated the sound as we walked around the room in circles. It was little moments like these that I came to delight in, making simple connections with these girls on a human level, not as guard and captor, or as enemies.

One day when I was helping Min-Jin learn some new English words and phrases, she said to me with a grin, "Peace out!"

I was surprised to hear such a colloquial expression come out of her mouth. "Right on," I replied, holding my fingers out in a peace sign. "Peace out!"

"I understand 'peace out' is the liquid from the penis, but what about the other thing, the nonliquid?" she asked.

I was perplexed. I had no clue what she was talking about. I enunciated the words "peace out" slowly, trying to see what other words she could be saying. "Oh! Piss out!" I exclaimed.

"Yes! Piss out!" she responded. "But what is the other thing. I think it is called 'shit'?"

I burst out laughing. "Where did you learn these words?" I asked.

"From the movie *Big Daddy*," she said. She told me she hadn't seen the movie but had read the screenplay in college. "We read movie scripts to learn English," she explained. I chuckled while I imagined a new generation of North Korean youth quoting Adam Sandler flicks.

Min-Jin also told me about a teacher she had in college who was very mean and strict. "We called him Bush," she said, laughing. "If we don't like someone we call them Bush after your former president."

One day Hyung-Yee came back from a break carrying a bunch of acacia flowers and leaves that she'd collected around the compound. The pungent aroma of the fuchsia-colored bouquet quickly filled the room. I breathed in deeply and closed my eyes, trying to take in as much of the sweet-smelling odor as possible. Hyung-Yee, spotting how affected I was by the presence of these beautiful flowers, plucked off a stem and handed it to me. I was grateful for this simple act of kindness and placed the stem beside my bed, hoping the fragrance would spread across the room.

Along with walking circles in my room, I did some basic yoga stretches. Every day, as I contorted my body or dropped to the ground in a pose, Min-Jin and Hyung-Yee would watch me, while making sure they weren't staring. One evening, when the power was out and a single flashlight illuminated the guards' area, they, along with two of the compound's women caretakers, quietly asked me to teach them some yoga moves. I was happy to oblige. I began with some deep-breathing exercises. Then I raised my arms high above my head and bent my body to one side. They followed in unison as we stretched first to the right and then to the left like trees swaying in the wind. I closed my eyes and thought about how nice it felt to be a part of something; that we were all sharing in one another's energy. Suddenly Hyung-Yee lost her balance and fell over. We broke into laughter. But this lightheartedness didn't last long. The mood quickly changed with Hyung-Yee's tumble, and the caretakers and guards assumed their usual formal and reserved personas. It was as if they had been caught doing something prohibited. They never asked me to show them any more poses.

The guards had a television in their quarters, and they watched it daily. I was told I could watch television whenever it was on, but I was often repelled by the continuous blare of military propaganda. The majority of the programs are black-and-white films that demonize the United States and lionize the North Korean regime during the Korean War. News segments are devoted to praising Kim Jong Il's leadership. The Dear Leader is regularly featured presiding over the opening of a new factory or the building of a school, and the screen is filled with shots of fertile fields, booming chicken factories, and military celebrations.

I became familiar with the Communist revolutionary songs and videos that played throughout the day. Sometimes Min-Jin would translate the meaning of the lyrics. If one of these songs spoke of love, it was most likely about love of country or for the common purpose of building a workers' paradise, but never about romantic love.

Most performers on TV sang along with an orchestra, violinist, or accordion player. The songs were traditional and classical in nature, in line with North Korea's conservative society. One especially popular musical group was a four-man acoustic guitar ensemble from the national army. While they sang the same songs I'd been hearing over and over, they did so with a folksy twang while strumming their guitars in unison. When the quartet appeared on television, dressed in their pristine green military uniforms and with perfectly coifed hair, the guards in the room would swoon. It seemed that even the strictest of cultures had its own boy bands.

The weekend programs offered slightly more variety, and I found myself looking forward to Sundays at 5:20 P.M., when two segments of the cartoon *Tom and Jerry* aired for a total of ten minutes. If I was in my room during this time, the guards would shout, "Miss Laura! *Tom and Jerry*!" It was the only show I could understand, because it has no dialogue. Tom just chases Jerry, day after day.

Watching the cat and mouse duo reminded me of a documentary a colleague of mine produced about the prisoners at Guantánamo Bay. Their favorite thing to watch was also *Tom and Jerry*. *How appropriate,* I thought. *The shenanigans between this mischievous cat and mouse seem to be a universal favorite for prisoners the world over, including myself.*

I also made it a point to watch an international news program that aired every Sunday at 7:30 P.M. Many of the reports focused on the U.S. military campaigns in Iraq and Afghanistan and the botched missions, such as a U.S. bomb that killed civilians. The declining U.S. economy and the fall of the Detroit auto industry were also highlighted. Wars, conflicts, and poverty in other countries were shown, as were natural disasters around the world. The North Korean media programmers seemed determined to show a world of fighting and chaos that was worse off than its own.

One night in early April, while I was eating dinner and Min-Jin was lying on the couch, her eyes half-shut during the local news, a report came on that caught her attention. I looked over at the screen and saw something blasting off into space. A Korean news commentator proudly narrated the sequence of events. Min-Jin sat up and glued her eyes to the television set, holding her hands to her mouth.

"What was that?" I asked.

She waved her hand in a hushing gesture, not wanting to miss a moment.

After watching the image replay for a third time, she turned to me beaming. "My country just launched a satellite. It is a very proud day."

My stomach turned. I knew this would most certainly complicate my situation.

"Wow, that's great," I lied. "You must be very happy."

I wiped up the area where I'd been eating, went into my room, and crawled into bed. I wondered how the U.S. government and the United Nations Security Council would react to this defiant act. I closed my eyes and tried to sleep as music from the North Korean military choir blared from the television set.

# CHAPTER FIVE

# the confession

⸰⸰⸰ LISA

O N APRIL 4, 2009, at approximately 10:30 P.M. eastern standard time, the media began reporting that North Korea had
just launched a long-range missile. We were gathered at my
mom's house, with our stomachs in our throats, as we listened to the
reports that North Korea had proclaimed its "peaceful" launch of
a satellite into orbit. The Obama administration and regional governments like Japan and South Korea immediately charged North
Korea with provocation and a violation of UN Security Council Resolution 1718, which expressly prohibits North Korea from
conducting ballistic missile–related activities of any kind. Many in
the global community were assailing North Korea for recklessness,
and the United Nations Security Council convened an emergency
session to figure out how to deal with its behavior.

UN Secretary General Ban Ki Moon, who is from South

Korea, said: "With this provocative act, North Korea has ignored its international obligations, rejected unequivocal calls for restraint, and further isolated itself from the community of nations. I urge North Korea to abide fully by the resolutions of the UN Security Council and to refrain from further provocative actions."

North Korea responded by asserting that it would pull out of the six-party talks if new sanctions were imposed on it. This meant that the country already considered the most isolated on the planet was threatening to turn inward even more. And for my family, this was happening while North Korea was holding my sister captive.

All these actions filled me with anxiety as I wondered what this would mean for the girls. Would this global denouncement make the North Korean government even more aggressive? In light of what was happening, we worried that the government might find more ways to use Laura and Euna as bargaining chips. We had to figure out how to separate the issues. We didn't want to entangle our situation with this nuclear conundrum.

This was a time when we were greatly comforted to have Governor Richardson as an adviser. In particular, he was a former U.S. ambassador to the United Nations, and he was able to provide valuable insights into what was going on behind the closed doors of the United Nations Security Council's emergency session. The governor was not overly alarmed by what was happening. He was certain that neither China nor Russia, countries friendly to North Korea, would ever agree to increase sanctions against it. He fervently believed that China's and Russia's protection of North Korea meant there would never be any real possibility of punishment by the UN body. This time, however, he was wrong.

On April 13, the fifteen members of the United Nations Security Council—including China and Russia—unanimously condemned North Korea's rocket launch as a violation of a UN resolution. However, though they called for the tightening of sanctions, they did not impose them.

Even so, this very public censure rankled North Korea enough that it declared a withdrawal from the six-party talks forever. A statement from Pyongyang proclaimed that it "will never again take part in such talks and will not be bound by any agreement reached at the talks."

North Korea immediately expelled nuclear inspectors from the country and also informed the International Atomic Energy Association that it would resume its nuclear weapons program. After that, North Korea went dark. Its leaders stopped talking to everyone, and Laura was somewhere inside.

Despite headline news stories about North Korea's nuclear tests, reporting about the girls was limited, in part because of the dire requests I'd made to news organizations to keep the profile low. Laura and Euna's company, Current TV, had also instructed all its employees to remain silent and pulled some of Laura's reports off the air and its Web site. But random individuals all over the country and the world were starting to ask questions over the Internet. "Why is there so little information about the American journalists?" some would ask.

A man in Philadelphia named Brendan Creamer, whom none of us knew, put up a page on Facebook called "Detained in North Korea: Laura Ling and Euna Lee." Thousands of supporters signed on, and Brendan provided telephone numbers and e-mail addresses of politicians, network news correspondents and executives, foreign countries' ambassadors to North Korea, as well as many other contacts. He posted updates almost daily demanding that news outlets cover the story and that politicians act to get these two American journalists out of North Korea.

Another person we didn't know, Richard Horgan, began a blog and Twitter site called "Liberate Laura" that tweeted daily any and all information related to the girls and North Korea. Liberate Laura's blogs and theories about why North Korea was holding Laura and Euna were fascinating to read and also attracted thousands of followers. On July 16, the Liberate Laura blog intimated that Song Taek, the man married to Kim Jong Il's younger sister, was running the country behind the scenes and was calling the shots in my sister and Euna's case.

He wrote:

> It's critical when dealing with a situation like that of Ling and Lee to know, or at least be able to visualize, your opponent. So think not of an ailing Kim Jong-il or his basketball-loving third son Kim Jong-un (even though the idea of a one-on-one game of high-political-stakes-B-ball between Jong-un and President Obama has a certain appeal). Rather, envision a lifelong, healthier politician who is said to be helping prepare Jong-un for the era of the "Dearly Departed Leader."

While Liberate Laura's hypothesis turned out to be untrue, it was an interesting read nonetheless. We later learned that Brendan had been a follower of my work for years, and Richard had just recently watched Laura's reporting from Mexico and was moved by it. These two independent sites ignited a movement that led to nationwide vigils and a petition to the North Korean government demanding the release of the girls. More than one hundred thousand people signed that petition within days of its going online. The social networks of Facebook and Twitter became an emotional crutch for me. In addition to Brendan's and Richard's posts, thousands of people—most of whom I'd never met—were regularly expressing messages of love and hope. These people became my world.

Whenever I was out in public, people asked me about Laura. "I don't know how she is," I solemnly admitted. "I don't even know where she is." It was the truth. At a certain point, I pretty much stopped going out, except for essential trips to a store or for meals. I felt an emptiness; my best friend was missing. I was tormented by voices in my head that wouldn't stop asking the same questions: "Where the hell is she?" "How is she?" My sister was out there alone and scared, and I couldn't pretend that everything was normal, so most nights I stayed home. Sometimes late at night, I'd post on Facebook or Twitter, "I miss her."

Instantly I would be deluged with messages from random people saying things like "My prayers are with you." Or "Stay strong, Lisa!"

The anonymity of these notes was strangely comforting. I didn't have to face people and all the unanswerable questions. Just knowing that so many people were sending positive energy in our direction uplifted my spirits during frequent periods of despair.

## ∿ LAURA

SOME DAYS, MR. YEE would show up without his recorder and notebook. This was a sign that we were going for a stroll outside. The first time he suggested we go for a walk, my suspicions were aroused. Breathing fresh air, feeling the cool breeze against my skin, hearing the rustling of leaves in the wind—these sensations had become rare and special to me. But the walks became more than just a time for exercise; they became an opportunity for Mr. Yee, Mr. Baek, and me to speak more freely with one another, with little chance of anyone overhearing us. I came to view these walks, our conversations, and the information gleaned as the most crucial part of my existence. I knew that if I was to ever get out, I needed to figure out how to play this complicated game with enormous international implications. I began to look at each day as a strategic puzzle, one I had to solve in order to win back my freedom.

As we strolled the length of the walled compound, I could see that Mr. Yee was trying to better understand my character, and I too had my own agenda. I wanted him to know me and sympathize with me. I tried to play the part of the naive young girl. I wanted him to see himself as my protector. At the same time, I was trying to extract as much information from him as he was from me.

"Do you consider yourself Chinese or American?" he asked during one of our walks. I knew which answer would please him, but I wanted to be honest and have an open conversation in this more relaxed setting.

"I feel a very strong connection to my Chinese culture and my heritage," I said. "But I was born in the United States, I was raised under its system, it's what I identify with."

He criticized the United States for its bullying nature and for thinking it can police the world.

"I agree with you," I said. "My country can definitely act like a bully. But our new president, Barack Obama, was elected in large part because he vowed to restore America's image in the world and to act as a partner, not as a superior, around the world."

He then started talking about sanctions and threatened that North Korea would retaliate if the United States continued to antagonize it. I wondered if this had anything to do with the recent satellite launch and America's reaction to it.

"Did something happen recently? Were more sanctions issued?" I asked.

Without answering the question directly, he went into a tirade about North Korea's right as a sovereign state to launch a satellite into space for peaceful, scientific purposes.

"Criticizing something that is our natural right is an insult and will only make us more determined," he said. "If the U.S. puts more sanctions on us, it only makes us more defiant."

He then told me that there were people within the regime that actually welcomed sanctions because they provided the government with a reason to rally the North Korean people against the United States.

While Mr. Yee became more candid and relaxed during the walks outside, when he was indoors he was all business. He approached the interrogation process as if it were some sort of duel, and he was always ready to pounce.

I began to tell from the questions being asked of me that the North Koreans were more focused on my work as a journalist than on the crossing of the border. This is a country where all news is censored and disseminated by the government through a strictly controlled propaganda machine; there's no tolerance for anything that deviates from the idealistic image the regime has created for itself. Mr. Yee explained to me that the North Korean government believes that the foreign media perpetuate lies about the DPRK's treatment of its citizens in order to crush the regime.

Perhaps the most damaging piece of evidence they had against me was the interview I did with a North Korean defector the night before we were arrested. Euna had tried to destroy the tape that contained this interview by ripping the ribbons when we were in the detention facility near the border. But apparently the authorities in Pyongyang had been able to piece together some sections of the tape and had seen at least part of the interview.

Unlike the women we'd interviewed, whose primary reasons for leaving North Korea were to find food and to make money to feed their families back home, this man had left for political reasons, less than two months earlier. He was disappointed with the regime and with the growing disparity between the ruling elites in Pyongyang and the rest of the impoverished country. I was curious to know if there were any underground movements in North Korea directed against the government, and whether others shared his frustration with the regime. He rarely answered me directly but indicated that he would explain more if he could escape to Seoul. He was plotting to make his way out of China to South Korea and was therefore trying to be extra cautious until he was in a safe place where he could speak more openly.

Euna was careful to film the man only from his waist down so that he would be unidentifiable. But the investigator didn't seem concerned about the man, his beliefs, or his identity. It was my questioning that most enraged him. He wanted to know why I had posed such political questions, including ones about Kim Jong Il's health and whether the average person was aware of the leader's condition.

"When we were introduced to the man, we were told that he left North Korea because he was unsatisfied with the government," I explained. "I asked him those questions only because I knew this. Otherwise, I wouldn't have asked him such political questions."

"I don't believe you," he responded angrily. "It's obvious you had another agenda."

It was soon clear that I was being cornered. The evidence they had on tape had convinced them I had hostile intentions. While I had apologized profusely for both crossing the North Korean border and for working on a documentary about defectors, my expression of regret was not sufficient for Mr. Yee. He wanted me to admit that my main objective was to bring down the North Korean regime.

That night, to my surprise, Mr. Yee, who as always was accompanied by Mr. Baek, visited me carrying a large bottle of beer. It was April 15, the birthday of North Korea's founder, Kim Il Sung. Throughout the week, different celebrations and films paying tribute to the Great Leader were featured on television. Mr. Yee ordered one of the guards to bring some glasses and filled three of them with the warm brew. Just then the room went dark, and the guard had to fumble around for a flashlight, which she placed at the center of the desk.

"Today is the great Kim Il Sung's birthday. It is a very special day for our country," Mr. Yee said excitedly.

Even though there was only a dim glow in the room, I could see that his cheeks were flushed, his speech slightly slurred. It seemed he'd already been celebrating. He raised his glass, and Mr. Baek and I followed along, clinking with each other in a toast.

I couldn't figure out what this visit was all about. He began by making casual conversation, asking me how my parents met and why they got divorced. I told him that it was a kind of arranged marriage, and because of that, it was hard for them to love each other.

"Divorce was probably the best thing that happened to me and my sister, because our parents fought a lot," I said. "I don't believe people should be together if they aren't in love."

"We do not approve of divorce here in the DPRK," he said. "It is something that is looked down upon." He then brought up Iain. "Your husband," he began, "he's quite handsome. He seems like he has good character, like he'd make a good Communist."

I was startled. How did he know what Iain looked like? What made him think Iain would be a good Communist? And what did that even mean? I wondered if perhaps another batch of letters had arrived from my family.

I saw that his glass was nearly empty. Trying to be deferential, I stood up and grasped the bottle with both hands, filling his glass first, then Mr. Baek's, and finally mine. He nodded in approval.

"How do you know what my husband looks like?" I asked.

"We have the best intelligence here," he answered proudly. "It's not hard to find out these things." He paused a moment, then said, "Miss Ling, do you think you have done enough in this investigation? Do you think you have confessed fully and frankly?"

"Well," I began, "I know I have tried my best to recall everything that I did leading up to my arrest. I know I am deeply sorry and regretful for my actions."

"I am not a powerful person," he said. "Do you know why I

never get promoted? Over the years, I have investigated many kinds of cases of people we've caught from Japan, China, and elsewhere. Every case I've had, I've managed to send these people back home."

I became cautiously excited at the mention of his sending foreign prisoners back home. I stared at him wide-eyed, and concentrated hard on whatever message he was trying to send.

"My bosses never promote me, because they want me to send criminals to prison. But instead, I end up seeing these people off at the airport."

I took a big gulp of beer and tried to process what he was telling me. I was almost giddy, thinking that he might be escorting me to an airport sometime soon.

Then his tone turned cold. "It disappoints me to hear that you think you have done enough. I think you have a long way to go," he said sharply. His words cut into me.

He called for the guard to bring over another bottle of beer. "I went to the Foreign Ministry earlier and read some of the reports coming out of your country," he said. "Even your own media is saying that according to their interpretation of DPRK law, you should receive at least fifteen years in prison. And that is your media! They know what a serious crime it is to be creating an anti-DPRK report about defectors."

The combination of his words and the alcohol was making me numb. I stared at him blankly, not knowing how to feel.

"But what is the use in keeping you here?" he said. "What is the use in you being here for much of your adult life? I don't think there's much use. So let's make a pact," he said, raising his glass once more. "You do your part, and I'll do my part."

I touched my glass to his, and he looked me in the eye.

"You need to use every ounce of your energy to think about your crime," he began. "What you are telling me is not going to be enough to get you home."

"Sir," I replied nervously, "I promise I will do better in the investigation, and I will do my part. I will make sure you do not get a promotion out of me."

He gave a loud chuckle, downed the rest of his beer, and got up and left.

I knew what he was getting at, and I knew that what he wanted from me was a confession. That night, I lay awake trying to weigh the implications of admitting to the colossal charge of trying to overthrow the North Korean regime. Would saying such a thing seal my fate and send me to a labor camp for the rest of my life, or might it pave the way toward forgiveness for what they believed was my true crime? Could I trust that Mr. Yee was being genuine and that he would do his part to get me home?

The next morning, Mr. Yee and Mr. Baek entered the room and Mr. Yee took his normal place at the desk in front of me. He lit up a cigarette, opened his red notebook, and pressed the RECORD button on his tape recorder.

"Well?" he said, taking a few puffs. "What do you have to tell me today?"

I sat in silence for a few moments, contemplating what I was about to say. Finally, I forced the words out, ever so slowly. I explained to him that I had thought long and hard about my crimes and that because of the perpetuation of stereotypes about North Korea's being a brutal country with no human rights, I decided I wanted to do my part to bring down its government. I admitted to having hostile intentions and to trying to topple the North Korean regime.

I didn't know if I was making the dumbest mistake of my life. I had confessed to the gravest possible crime and handed him everything he needed to send me to the firing squad. Had I just walked into a trap from which I might never escape? But there was no turning back. As Mr. Yee took down notes and the tape recorder captured my every utterance, I could only hope that I'd made the right decision.

Once I had confessed, Mr. Yee's attitude toward me changed dramatically. He seemed satisfied with my answers. The air became less tense, and Mr. Yee appeared to be more relaxed. I think the most important part of his work—getting me to admit to having hostile motives—had been achieved. It had been more than a month since we'd been captured, and it was still unclear what was going to happen to us.

☙ LISA

Because I was the older one, I suffered the punishments and groundings from our dad first. Of course, looking back, I suppose I deserved to be disciplined for stealing the car at age fifteen and hiding Dad's beers in my dresser drawers. Growing up in a community with few Asians and being the older sibling, I always needed to feel that I fit in. At school, I was more concerned with popularity than geometry, and that was clear in my grades. I just didn't care much; I wanted to have fun. My sixth-grade teacher, Mr. Smith, called my father in to a meeting specifically to discuss my behavior. He told my dad that I was too much of a "social butterfly" and that my gregariousness was affecting my performance in school. My father had to take time off from work to attend this meeting, and he promptly grounded me for a month. But that didn't change my ways. My efforts at becoming popular were fairly successful. I was determined to transcend the fact that I had slanted eyes and that our house always smelled like Chinese food.

I was such a rabble-rouser that by the time Laura reached her teen years, Dad had mellowed a bit. By this time, our grandmother was in a convalescent home and our dad was the only adult in the house. Assuming responsibility for two teenage girls was not an easy task for a man, especially one who had to work such long hours. I think I wore him out with my mischievous behavior. Compared with me, Laura was his little angel. She was quiet and sweet and excelled in school. She was the smarty while I was the chatty Cathy. I was lucky if I figured out how to get more than one A on my report card, while Laura never got anything but As. With our mom in L.A. and our dad working at night, I was the one who went to a number of Laura's open-house events. The test scores were posted on the wall of her classroom, and hers were either the highest or the second highest in the whole class. This made me really proud.

Once, Laura complained to me about a teacher who was being unfair to her. The next day I marched into the principal's office and demanded that she be transferred to another class. The following week, it was done. Some siblings are competitive with one another, but in some ways I felt more maternal toward my sister; she was like my kid.

During my junior year of high school, I was asked to the prom by Ren Bates, a boy I had liked for a long time. I was so excited. My dad had given me some money, and I had gone to the mall and bought the coolest purple sequined dress. It was so different from the froufrou, lacy gowns most girls picked. It was the most beautiful thing I owned. On the night of the prom, couples typically went to each other's homes and the eager parents took zillions of photographs.

I spent hours curling my hair and putting on my makeup. I bought a new magenta lipstick for the occasion. Then I put on the dress. I loved it. I hoped my dad might surprise me by coming home from work early. I stood anxiously at the window, watching for any signs of his car. It never came. Instead, Ren's Chevrolet pulled into our driveway. When he knocked on the door, I could hear Laura's little feet run down the stairs; she was holding a camera. I opened the door and Ren came inside. My twelve-year-old sister motioned us to the living room and positioned us in front of the sofa.

"Stand up straight," she proclaimed with her braces-laden teeth. "Say 'cheese'!"

Laura knew this was one of the most important days of my life, and she did not want me to be let down. As Ren walked me out the door, I turned around to see this little person furiously snapping more photos of us as we approached the car. I will never forget that day: my baby girl came through for me.

~⌒ **LAURA**

W HILE I STRUGGLED HARD to keep my spirits up and to avoid falling into a depression, there were dark, lonesome days when I faltered. Some days, I'd retreat to the bathroom, look at myself in the mirror, and fall into a delirium. I would talk into the mirror as if I were speaking to another person, not myself. "Who are you, and what did you do with Laura?" I would say out loud, though quietly enough that the guards could not hear. I scolded the person looking back at me for letting her life slip out of control. Unable to contain my anger, I would flush the toilet hoping to create a noise that was loud enough to hide the sound of me slapping myself continuously until I was red in the face. I wanted to punish myself for hurting my family. At the same time, the stinging pain made me feel more alive.

A couple of weeks after I was given the first letter from my family, Mr. Yee showed up with another manila envelope. I wanted to grab it out of his hands and rip it open at that very moment. Inside were letters from Lisa, Iain, my parents, relatives, friends, and colleagues. Some people had written to me several times. The dates on them indicated they were sent at least two weeks earlier. Apparently they were sent via e-mail to the U.S. State Department, which then screened them and sent them on to Ambassador Foyer, who delivered them to North Korea's Foreign Ministry, which I'm sure looked them over as well.

I immediately went for the ones from Lisa and Iain and scanned through them for any information. It seemed the North Korean authorities had the same idea—Lisa's and Iain's letters were at the top of the pile, organized by date. Some of the pages even had coffee stains on them. I wondered whose eyes had perused the letters before mine.

I buried myself in each page and studied every word for a sign that something was happening that would get us home. But even though the letters were vague about any efforts that were under way, they were full of love and encouragement, things I needed most during my fits of desolation.

In one letter from Lisa, I could see that she was strategically sending messages of apology and goodwill to the North Korean authorities. She wrote:

> I hope those who are holding you know that you and
> Euna didn't mean any harm. Maybe this incident will
> provide an opportunity to establish a better relationship
> between our two countries. At the end of the day, we're all
> human beings and the main reason there's ever hostility is
> because there's been so little face to face: people don't take
> the time to get to know one another and instead believe the
> sensationalism that is proliferated with such constancy.

Reading her words made me proud. Here was my sister trying to affect the situation even though she was half a world away. But while Lisa was being calculating in her words to the North Koreans, at the same time she was also being a normal sister by cheering me up with frivolous details. She told me our dad had finally let her tweeze his gray eyebrows, something I'd been begging him to let me do forever. "I told him to do it for Laura!" she wrote. "And he did. I must say he looks infinitely better." I laughed out loud imagining Lisa plucking out the gray hairs of our curmudgeon of a father.

She also told me she got a tattoo. She'd been cooped up at our mom's house and was so exhausted from worry that she decided to do something to take her mind off things. "The tat says 'love and peace' in Arabic, and it's in the shape of a dove. I actually love it and can't wait for you to see it. I'm gonna try to convince you to get one too," she wrote.

That night I lay in bed thinking about what tattoo I might get. "I hate the DPRK" and "FU*NK" were just a couple that came to mind.

SOME DAYS DURING OUR walks, Mr. Yee and I would make small talk. He asked me what I liked to do on the weekends back home, or how often I ate out at restaurants. I told him I liked to go to the beach and that my husband liked to surf.

"Is surfing like in that movie where there is a gang of bank robbers who put on masks of the various U.S. presidents? They were surfers, right?" he asked, referring to the movie *Point Break* starring Keanu Reeves and Patrick Swayze.

"Exactly!" I said, astonished he'd seen the film. He told me he also liked the 007 James Bond flicks, though he preferred the Sean Connery versions to the newer releases.

"We also like to go to the ocean," he said. He explained that sometimes in the summertime after work, he and his colleagues went to the beach and barbecued clams that they found on the sand.

"That sounds like a lot of fun," I replied.

It felt strange to imagine Mr. Yee in his life outside of the investigation. I never asked about his personal life, and he rarely brought it up. Once, in passing, he mentioned having a wife and kid, but beyond that, his world was a mystery to me.

Every few weeks Mr. Yee came bearing another envelope of letters. I heard from close friends and relatives as well as people I hadn't spoken with in years. Many updated me on their day-to-day activities. It made me feel close to them, even though I was locked away on the other side of the planet. Some friends sent meditation techniques, which I immediately put into practice. My mother included a mental lesson each day. Others sent poetry, jokes, trivia, and puzzles. One poem, Maya Angelou's "Caged Bird," came to mean a great deal to me and I began reading it every morning. I derived strength from reading Angelou's words about the caged bird singing for freedom, not because I thought of myself as a caged bird, but because it made me think of the brave souls of North Korea, caged in a political system that was denying them their basic freedoms. I reflected back on my interviews with the defectors who had fled North Korea in search of better lives.

EVENTUALLY I DEVELOPED a sense of trust with my guards, and they became more relaxed and lenient with me. During the day, they would partially open the curtains in their room to let in some natural light. They also let me sit in their area, and as long as I kept a safe distance from the window, I was allowed to look outside and take in the scenery. I developed a daily routine: after I finished my morning duties of mopping the floor, dusting off the furniture, and cleaning the bathroom, I'd sit and peer out the window. Over time, I observed small leaves form on naked tree branches, and I watched as a blackbird searched for twigs to build a nest. It was strange to be witnessing nature and the changing of seasons from behind a window, without being able to inhale the fragrance of a blossoming flower or to feel the day's first dew on shoots of grass that were just emerging.

I never brought up politics or my case with the guards, but there were times when they initiated conversations about their political system.

"We are not a rich country," Min-Jin said one evening while I was eating dinner. "But we have our pride. The United States puts sanctions on us because they don't agree with our system. Why do they need to pick on us? We are just a small country. Our leader, Chairman Kim Jong Il, has worked so hard to provide for us, even though the rest of the world tries to hold us back. We have never acted aggressively toward any country. It is other countries that have always invaded our land and attacked us. If they attack us, we will fight."

As she spoke, her eyes started to well up with tears. "Our leader is getting old, and it hurts me to see him get old because he has devoted everything to us. We want to work hard to make him proud."

I told her I believed our two countries could become friends in time and that perhaps things might change with the election of our leader, President Obama.

"Every time you elect a new president, you say things are going to be different. But nothing ever changes," she said. "I don't think our countries will be friends in my lifetime."

Min-Jin's sentiments did not surprise me. Her feelings seemed to stem not only from an intricate propaganda network created to instill hatred for the United States in the North Korean people, but also from a history of mistrust between our two countries. I knew that her opinions were not hers alone but were shared by the North Korean people and government, and I feared they would want to punish me as a way of hurting the United States.

O N APRIL 24, NORTH KOREA finally broke its silence and issued a statement that, of all things, pertained to Laura and Euna. A trial date had been set. According to the Korean Central News Agency (KCNA), "A competent organ of the DPRK (Democratic People's Republic of Korea) concluded the investigation into the journalists of the United States."

The trial date was set for June 4. I had just returned to my mom's house after having lunch with a friend when the news broke online. Mom and I sat staring at our computer screen unable to speak for a few minutes. What did this mean? In previous cases of Americans detained in North Korea, releases were obtained after some diplomatic wrangling. This was unprecedented—no American had ever been tried in North Korea's Supreme Court.

The phone rang. It was Iain. His voice sounded shaky. He was obviously distressed by the news. He said he was coming over. Iain is a quantitative analyst for an investment fund, and he went back to work a couple of weeks after Laura was arrested. His bosses could not have been more understanding of what he was going through. They said he could take all the time he needed and they meant it. Since most of our days were filled with sitting around and waiting, we told Iain he should go back to work as a way of settling his mind. From his office, he shot me about ten articles a day about anything related to North Korea, and we spoke by phone multiple times throughout the work hours. Iain was devastated. In twelve years of knowing him, I had never witnessed him emotional or even the least bit shaken. Although he tried hard to maintain his composure, I saw a side of Iain that was deeply painful to behold. He started losing a lot of weight, and I would see his eyes well up in tears as he stared off into space. Laura was his world, and she had suddenly disappeared.

I phoned Governor Richardson, who told me not to overreact. "This date of a trial is good news," he said. "At least they [the North Korean government] have given a firm date. Hopefully I'll be able to go over there soon after that."

We would have to wait until June 4 for North Korea to make its next move. In publicly setting this time frame, the North Korean government seemed to be declaring that there would not be an immediate release. Laura and Euna had already been detained for more than a month. And June 4 was more than another month away. Every day without word about my sister was a day too long.

## ∾ᴄ᷉ LAURA

THOUGH I'M THE YOUNGER sibling by three years, I've always been the mature one. Lisa calls me Baby Girl, but it's really she who is the more childlike of the two of us. Our dad often refers to Lisa as the "Senior Teenager" because she knows the name of every pop song on the radio. Our parents actually took a stab at being a nuclear family when Lisa was first born, which allowed her to have some sort of childhood. She was their first kid, after all, and perhaps they saw this as a chance for a new start in their rocky marriage. When she was really young, our parents doted on her, threw elaborate birthday parties for her, and took her to ballet and tap dance classes. But this charade of trying to be the perfect parents didn't last long, and the fighting carried on.

By the time I arrived, it was clear my parents' marriage wasn't meant to be. I can't recall a moment when Mom and Dad were happy or laughing together. No photos exist from my dance recitals, because I never had any. My mother's nickname for me was Yun-tsai, which in Chinese means "little old person." I never made a big deal of my birthday, because unlike the other kids, who had pool and pizza parties, I always knew my birthday would just come and go. That changed one year when Lisa, realizing my birthday was quickly approaching, decided to throw me a little surprise celebration. I was turning ten and she was thirteen. She phoned up a few of my friends the night before my birthday and asked if they could come over to our house after school the next day. When I walked into our living room, I was greeted by a group of classmates screaming "Surprise!" Lisa was standing in the middle of the room with a look of pure glee. It wasn't an extravagant setup, just a few friends playing games. But I'll never forget when we gathered around our dining room table, and Lisa entered the room carrying a chocolate cake she'd baked with the help of my grandmother. She had the biggest smile on her face. To this day, I don't like to celebrate my birthday. But no matter what, Lisa never lets my birthday pass without at least organizing a dinner with my close friends.

As I sat in the guards' area looking out the window, I thought back to that cheery December afternoon when I didn't have a care in the world. While I didn't know what would become of me, for a brief moment I felt free.

Reading letters became another escape for me. I was grateful when Mr. Yee began to give me batches of letters on a fairly regular basis; they usually came every two weeks, but sometimes I would get them weekly. Some of the packets were thicker than others. But the one constant was hearing from Iain. One batch of letters might contain five letters from him, each several pages long.

My husband is the unflappable one of us, while I make my emotions known. Among friends, Iain always has a sunny disposition, but it's hard to penetrate what he's actually feeling. I think I'm the only person he's confided in and opened up to for most of his life. In Iain's first few letters to me, he seemed reserved; his tone was very pragmatic. I knew he was having trouble expressing himself on paper. He phlegmatically wrote about his days at work, updated me on news headlines, and offered words of encouragement. In one note he wrote, "I can't always express what I'm feeling, but I know you know what I mean." I did.

But as time went on, Iain's words became more emotive. He began to pour out his heart to me, even though he knew his letters were being read by multiple staffers at the U.S. State Department and countless members of the North Korean regime. He wrote to me two times a day, and sent both typed and scanned handwritten letters so that I could see his penmanship and feel closer to him.

Every day at 5:00 P.M., Iain wrote to me from our dining room table in Los Angeles. He scanned a photo of himself to show me where he was situated so that I could think about him in that familiar location every day. This became our virtual meeting time. At 9:00 A.M. in Pyongyang, which was 5:00 P.M. in California, I sat in the guards' area, looked out the window to the sky, and envisioned Iain writing to me. It became our time together, and during my captivity he never missed a date. His letters were a source of strength, comfort, and love. They were my oxygen.

In the letters, Iain included news summaries from the *Economist* and the *New York Times* to keep me up to speed on what was happening in the world. He scanned sheets from a journal I wasn't even aware of, one he had kept more than a decade ago when we first started dating.

One journal entry was dated May 11, 1997, the day after we met at the concert. Just seeing his handwriting from that time brought me to tears. He wrote:

> *When I first saw her, I went "Wow!" I chat to her a bit, but nothing else. Then she rings today—we chat for about an hour and arrange to meet for coffee, go to the Geffen*

*[Museum] when I get back from Miami. What am I going to do? She is beautiful.*

Iain also included scanned photos from our vacations together, parties, our wedding, and family gatherings. Whenever there was a picture, Min-Jin and Hyung-Yee gathered around and looked over my shoulder curiously. They giggled like teenage girls at the sight of one of our wedding photos.

"You look very beautiful!" Min-Jin said, translating for Hyung-Yee.

"Yes, you look different than you do now. Much better with makeup," chimed in Min-Jin.

I laughed. I remembered what I had looked like when I first met the guards. Battered and bruised, I resembled a wild animal. While my wounds had healed externally, I was still straggly and unkempt, and continued to feel pain on parts of my head and face.

In one photo that showed me wearing a long halter dress that exposed my bare shoulders, Min-Jin commented, "Wow, your dress is so sexy!" She had on a disapproving look. I hardly considered the full-length dress seductive, but in North Korea anything showing the slightest bit of skin, even just the shoulders, is considered unacceptable.

"Would you ever wear something like this?" I asked Min-Jin.

"No. It is not customary in our culture to wear such things. We also don't wear blue jeans, because they are a symbol of America. But you are American, so it looks good on you," she said.

About two months into my captivity, when it seemed the investigation was winding down, I pleaded with Mr. Yee to let me have one book that had been sent by Ambassador Foyer early on. Later that day, he came to my room carrying Ian McEwan's *Atonement*. I devoured every word. I would close my eyes after reading passages that described the English countryside and imagine that I could feel "the cool high shade of the woods" and see "the sculpted intricacies of the tree trunks." I fashioned McEwan's fictional world in my mind so vividly that I could almost smell the rhododendrons. The novel depicts two lovers, Robbie and Cecilia, who become separated by war. Despite Robbie's determination to make it back to England to be with Cecilia, he dies in France before he can be reunited with his love. It was hard to keep from imagining that the same fate could be mine, and that I would never see Iain again.

Along with the letters, I began receiving some packages from family and friends back home. One of the first parcels I received was from a dear family friend, Morgan Wandell, who lovingly sent some much-needed items such as tweezers and nail clippers. While I was able to keep these basic things, I was denied the granola bars, chewing gum, and playing cards that were also included. I had stopped caring about my appearance or personal hygiene, but I was excited to get these small items, which helped occupy my time. Rather than immediately cutting my nails, which had grown long and chipped, I decided to wait until the weekend, to give me something to look forward to. These mundane acts would become another source of escape throughout my captivity.

In Iain's letters, he said that he was sending one small package a week, but I normally received a batch of items every few weeks. Because of the sanctions, it's almost impossible to send parcels to North Korea, so Iain would send them to his family in London, and they mailed them to Ambassador Foyer in Pyongyang. Most of the packages contained things like clothes, toiletries, books, magazines, and food. Ironically, my family sent me packets of dried seaweed, one of my favorite snacks. It's a product of South Korea that is imported to the United States. Now it was being sent back around the world to North Korea. I wasn't allowed most of the food, but after persistent begging and complaining that I needed certain items for my ulcer, I was given some crackers, seaweed, and dried fruit. I rationed these delicacies, allowing myself one or two pieces each day.

In one parcel Iain included a cashmere sweater he'd given me one birthday, one of his T-shirts so that I could feel closer to him, and a bottle of shampoo from a hotel in Napa Valley where we'd vacationed one summer. I washed my hair with it the next morning, and the sweet fragrance of blood oranges transported me back to Northern California's rolling hills and wine country. I slept with his T-shirt beside me each night, putting part of it up to my nose. When I first received it, it smelled so much like my husband that some nights I'd wake up thinking he was right beside me. I'd open my eyes to see the same drab wall, but at least I had a piece of Iain with me.

Now, every few days, I was allowed a new book, and I pored through each one. They were my escape into other worlds. I seemed to find a parallel to my own situation in nearly every novel. In *The Adventures of Huckleberry Finn,* Huck reflects on the boredom of living under the care of the strict widow, Miss Watson. "Then I set down in a chair by the window and tried to think of something cheerful, but it warn't no use. I felt so lonesome I most wished I was dead." As I peered out the window in the guards' room, I felt Huck's despair.

Along with books, I was allowed to read a few magazines at a time. Family and friends had included an array of publications including the *New Yorker, Vanity Fair, People, Glamour,* and *InStyle.* I'd never been as up to speed on the latest Hollywood celebrity gossip as I was during my time in captivity. My guards were interested in the fashion magazines. One day, Min-Jin asked if she could read the *Glamour.* "Of course," I said and handed it to her. As she was flipping through the pages, Hyung-Yee walked into the room and began looking on as well. They glanced over at me and put their fingers over their mouths, indicating that I shouldn't tell anyone they were reading the fashion magazine. I immediately became nervous. I had no problem giving anything to my guards if they asked for it; they were my guards after all. But I was not about to do anything that could get me in any more trouble than I already was. At the same time, I didn't want to fracture the bond I had worked so hard to build with them. I decided my relationship with my guards was less important than jeopardizing my already dire situation.

I waited anxiously for the girls to finish perusing the magazine and then said to them, "I'm sorry to say this, but I can't let you look at these magazines if you don't have permission. I can't do anything that might cause any more problems for me. I really need to get home to see my family. I hope you understand." They nodded in acknowledgment.

"We understand," said Min-Jin. "It's no problem. We just wanted to have a quick look."

"Okay, great," I said, relieved that I hadn't burned a bridge with them.

One evening, while watching the nightly news report, I saw an image of Roxana Saberi, the Iranian-American reporter who had been arrested in January while living and working in Iran. I was familiar with Roxana's case because it became international news after Saberi was detained for allegedly purchasing a bottle of wine in Iran. My North Korean guards were fixated on the report. After the newscaster finished, there was an awkward silence in the room. The guards whispered to each other while stealing glances at me. I asked what the report was about. One of them explained that Saberi had just been sentenced to eight years in prison for espionage. The news about Saberi put me in a daze. I suspected that the North Koreans were paying attention to her case, and might even look to her situation as a model for ours. Feeling defeated, I retreated to my room in tears as the guards continued to murmur to each other.

EACH TIME MR. YEE took me outdoors for a walk, I asked him if he'd heard any news or knew of any progress.

"Your government doesn't care about you," he said to me one day. "They are trying to keep things quiet. You're not important enough to them."

"You're right," I agreed. "They don't care about us. We don't work for the government. We're just private citizens. The government is not going to get involved in our case."

I wanted to temper whatever expectations existed about our value. I knew that if the North Koreans were hoping for a big prize from the U.S. government in exchange for our release, the chances of our going home would be slim. I was hoping some sort of private monetary exchange might be sufficient.

"I think my family could raise some money so that they could pay the government a fine for our crime," I suggested. "Perhaps they could raise one or two hundred thousand dollars."

Again I was trying to reduce their expectations by suggesting an amount that wasn't overly exorbitant.

"We don't care about money," he responded. "That is what you value in your country—money and possessions. Here, things are different. Your government must do something. Our countries are at war, and you are a citizen of the United States. Your government must act."

He indicated that in the case of another American detainee, Evan Hunziker, who was arrested while swimming in the Yalu River along the Chinese–North Korean border, the U.S. government sent an envoy, then congressman Bill Richardson.

"What if the chairman of my company, former Vice President Al Gore, came as an envoy," I suggested. "Would that be acceptable?"

"Al Gore is a private citizen now. He's not connected to the government," he replied.

"Yes, but he is close to President Obama," I explained. "He's also a Nobel Peace Prize winner. He's probably one of the most popular postpolitical figures in the entire world. And I know he would be willing to come here."

He thought about this as we walked along. "That might be a good idea," he said, nodding.

Finally, he'd given me a response that offered a tangible way forward. I felt relatively confident that Al Gore had already offered himself as an envoy, and was continuing to do so. But I wanted to be sure.

Ever since my meeting with Ambassador Foyer, when the ambassador told me that I had a right to send letters, I'd been asking Mr. Yee if I could write to my family. "I will think about it" was always his response. I tried to persuade him that I could use my letters to express the need for urgent action, and then my government would understand the gravity of our situation.

During one of our walks, Mr. Yee asked to whom I would write, if I were allowed to send out letters. I told him I wanted to write letters to my family, to my sister, to my husband, to my boss and CEO at Current TV, and to my colleagues. He asked what I planned to write to each one of them. I discussed what I wanted to say in general terms. Strategically, I wanted to convey specific messages to my sister, who had contacts in the United States that could be helpful to our situation, and to my bosses at Current TV. I also hoped to write to my family, husband, and colleagues. I wanted to comfort them and let them know I was hanging in there and being as strong as possible.

"You must express the need to act swiftly," he chimed in. "If you go to trial, you will undoubtedly be given a very long sentence because of the nature of your crimes."

I knew his message had to be part of the government's overall plan. If he was telling me to pressure my family to act quickly before a trial, he must truly believe or know there was some sort of opening for negotiation. I took this as a positive sign, despite the growing tension between the United States and North Korea.

I was allowed to write the letters in mid-April, but it would be a month before I was able to give them to Ambassador Foyer during our second meeting.

⟨⟩ LISA

O N MAY 11, MY mom, Iain, Paul, and I flew to Washington, D.C., to meet first with Secretary of State Hillary Clinton and then with the Chinese ambassador to the United States, Zhou Wenzhong.

Even though China had not been active on our issue, three months had elapsed and we figured we had nothing to lose by asking for the Chinese government's help. Euna's husband, Michael, and daughter, Hana, joined us for the meeting with the secretary of state. The secretary's office also invited former Vice President Gore to attend because of his role as chairman of Laura and Euna's employer, Current TV. Plus, he had obviously known the secretary of state for a long time, and she thought it would be a good idea to have him in the meeting.

In her stately office, with Deputy Assistant Secretary Jim Steinberg as well as Kurt Tong, Linda McFadyen, and a number of other official-looking State Department colleagues in tow, Secretary Clinton expressed her concern about Laura and Euna's detainment. She told us that getting the girls back was important to her, both as America's chief diplomat and as the mother of a daughter. We felt consoled by how compassionate she was. There was a softness about her that doesn't always come through on television, and for that we were grateful.

Secretary Clinton spelled out the complications of dealing with North Korea—particularly because some of its neighboring nations were insisting on punishment for its recent aggressive actions.

"This is a uniquely American issue," I urged. "Surely, other

All-American girls with our all-American dolls.

This portrait of our paternal grandparents, Lien and H.T. Ling, was taken in 1946 in Nanjing, China.

This was a trip to a temple in Taipei, Taiwan, in 1948 to celebrate the birth of Black Dragon's second son to his second wife *(right)*. Our grandmother, Mrs. Wang *(left)*, is carrying our mother.

This photograph was taken in 1953 at a sixtieth birthday gathering for our great-grandmother *(seated in the center)*. Our maternal grandfather, known as Black Dragon, is in the center wearing the dark shirt. Our grandmother, his first wife, is standing next to him. Next to her are his concubines. The man next to Black Dragon is his brother and next to him are his two wives. Most of the children in this picture belong to the two brothers. Our mother stands just below Black Dragon.

Our parents, Douglas and Mary, on their wedding day, March 8, 1969.

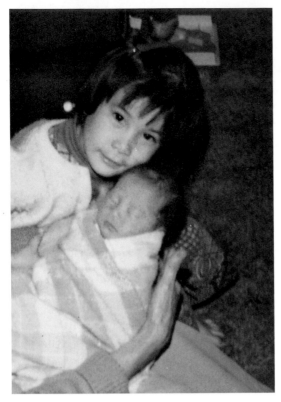

This is a photo of Laura just a couple of weeks after she was born in December 1976. Lisa was so excited to have a baby sister, or "Baby Girl," as she would call Laura.

Here we are with our grandmother, Lien Ling, who helped raise us, and made us memorize every church hymnal in the book.

Dad and his girls in 1977.

Here we are with our mother in 1980.

Laura and Iain dated for seven years before finally tying the knot on June 26, 2004.

Laura and Iain having fun with their bridal party in the courtyard at the Pacific Asia Museum in Pasadena, California. From left to right: Dorothy Fong, Michael Tummings, Laura, Iain, Lisa, Jean Roh, Josh Clayton, and Charles Clayton.

At Lisa's wedding on May 26, 2007. You'd never know our parents were divorced from this family photo. They were even color-coordinated.

Laura and Doug celebrating Lisa's wedding.

In June 2007, Lisa went to North Korea with Dr. Sanduk Ruit, where he performed cataract operations. Here, hundreds of people are lined up in Pyongsong, North Korea, to get checked by Dr. Ruit and his team.

In much of the world, cataracts occur in elderly patients. Due to severe conditions, many North Korean children are stricken with it. This is a young North Korean child who had just undergone a cataract operation on his right eye.

After a successful cataract operation performed by Dr. Ruit's team, this woman rushed to the portraits of the Great and Dear Leaders to thank them for restoring her vision.

Lisa and a man she asked to take a picture with in a park in Sariwon, North Korea. He is wearing a pin with a portrait of North Korea's Great Leader, Kim Il Sung. The documentary Lisa made after this trip was highly critical of the North Korean regime.

Laura's last assignment before going to China was for a Current TV documentary about the drug wars in Mexico. Here she is with producer Mitch Koss, who is filming her in front of a military base in Tijuana.

Lisa and Laura on New Year's Day, two and a half months before Laura's arrest.

In June 2009, Iain spoke at a vigil held at the San Francisco Academy of Art.

Iain sent dozens of letters and photographs to Laura while she was in captivity. In this one, he is holding a special message for Laura (he calls her "b").

Lisa and her husband, Paul Song, at a Los Angeles vigil for Laura and Euna.

This vigil in support of bringing Laura and Euna home was held in Los Angeles on June 4, 2009. It was one of a number of gatherings that took place the day Laura and Euna's trial was to begin.

Anderson Cooper's *360* broadcast live from this vigil, where Lisa, Euna's husband, Michael Saldate, and Iain appeared. It was believed that North Korea's Dear Leader watched CNN, so this live shot was orchestrated to happen the morning of Laura and Euna's trial in the hope that he would be watching.

Our father spoke at a vigil on July 9 in our hometown of Sacramento, on the steps of the capitol. The next day, Secretary of State Hillary Clinton asked the government of North Korea to grant amnesty to Laura and Euna.

This photograph, taken on August 4, 2009, of the jubilant North Korean leader, Kim Jong Il, and a stoic President Bill Clinton in Pyongyang was seen all over the world.

Iain, Lisa, and Mary holding a December 7, 2007, copy of *GQ* magazine on which President Bill Clinton graced the cover (not sure why we saved this old issue). They had just learned that Laura and Euna were safely on the plane headed home. Lisa e-mailed the photo to Doug Band, who showed it to President Clinton while on the plane.

This photograph, taken by President Clinton's close adviser, Doug Band, is of the vehicle carrying Laura and Euna to the airport in Pyongyang. Band wanted to make sure the car with Laura and Euna was never out of his sight.

While the plane taxied on the runway, Euna and Laura looked through the cockpit to see their families anxiously awaiting their arrival.

Just moments before the plane landed in Burbank, California, on August 5, 2009, Laura and Euna posed for this photograph with President Clinton and his team. From left to right: Justin Cooper, Dr. Roger Band, President Bill Clinton, Euna Lee, Laura Ling, Min Ji Kwon, Doug Band, David Straub, and John Podesta.

Mary hugging Laura moments after her return, while Iain and Lisa look on.

From left to right: Charles Clayton, Vice President Al Gore, Iain, Laura, President Bill Clinton, Lisa, Paul Song, Mary, and Doug.

In September 2009, our family threw a luncheon at Woo Lae Oak restaurant in Arlington, Virginia, to thank Secretary of State Hillary Clinton and State Department colleagues who helped with our case.

Kurt Tong (*left*) and Linda McFayden (*center*) of the U.S. Department of State were our regular government contacts and friends during Laura's captivity. We were so happy that Swedish ambassador to North Korea, Mats Foyer (*right*), happened to be visiting the United States when we threw the luncheon and was also able to attend. Ambassador Foyer was Laura's lifeline to her family and the outside world.

countries should understand that this has nothing to do with the larger geopolitical issues."

Without explaining how, Deputy Assistant Secretary Steinberg said that overtures had been made to North Korea to have direct talks about Laura and Euna, but there had been no response. Secretary Clinton reiterated the importance of maintaining low visibility so as not to raise the stakes and possibly provoke the people who were holding Laura and Euna. She strongly suggested that we continue not speaking publicly about the matter.

Then Al Gore chimed in. He told Secretary Clinton that he would be willing to go to North Korea if the opportunity presented itself.

To that, she remarked, "That's not a bad idea, Al. That just might work."

I was worried about what this would mean for Governor Richardson. No one in the room knew I had been communicating with him. If Gore were appointed to this mission, would the governor be out? I had only one objective, but I wondered if a change might result in the bruising of egos. The governor had become a confidant, and I knew he took our case very seriously. Still, at the end of the day, all I cared about was getting my sister out, no matter who was taking it on. I even thought the more people trying, the better.

That same day, we went to see Ambassador Zhou at the newly built Chinese Embassy in Washington. Designed by the two sons of renowned architect I. M. Pei, this structure is an imposingly modern, sleek symbol of China's emergence. Even though Governor Richardson had told me that the North Koreans loathed having to answer to China, we hoped that given our Chinese descent, the Chinese government might help us somehow. We presented Ambassador Zhou with a Hermès tie upon entering the room. He was formal but kind enough.

We told him we believed the North Korean soldiers crossed

into China to apprehend Laura and Euna. We asked if the Chinese government would raise this with the government of North Korea.

"I'm sorry but we cannot help you," he replied. "First of all, they were inside China without the appropriate visa."

What the ambassador said was true. Laura and her team had gone to China as tourists instead of journalists because they wanted to avoid the watchful eyes of Chinese government officials while reporting on the controversial issue of trafficking. In other words, he made it clear that we shouldn't look to China for help. That said, he kept the tie.

Meanwhile, Al Gore became a man on a mission. He had been engaged from the start, but after our meeting with Secretary Clinton, he went from taking cues from the State Department to becoming an active player in the game North Korea was perpetrating. Getting the girls out was among the most important things on his list of priorities. Additionally, the State Department had apparently found another means of communicating with North Korea outside of the "New York channel," and Gore allowed himself to be presented as an envoy to this other source. He was certainly a formidable candidate: a former vice president of the United States, former presidential candidate, Nobel Peace Prize winner, and chairman of the company that employed Laura and Euna.

We were starting to feel more momentum than we'd had in months. Gore brought his former national security adviser, Leon Fuerth, on board to be his lead in diplomatic efforts. Things were looking good.

Then, entirely unexpectedly, we were thrown for a loop. No more than a few days after our meeting with her, during an official press conference with Malaysia's foreign minister, Secretary Clinton made the following public statement regarding North Korea's designation of a June 4 trial date for Laura and Euna, when asked about it by a journalist in the room: "Actually the trial date being set we view as a welcome time frame," she said. "We believe that the charges are baseless and should not have been brought, and that these two young women should be released immediately."

The statement was picked up by press all over the world, and immediately confusion set in. We had just been in the secretary's office discussing the importance of not antagonizing North Korea's leadership, and I didn't know where the idea that the charges were baseless came from. The secretary was calling into question North Korea's legal system. Since words mean everything in North Korea, we wondered what this might do to our case.

Our family had been ultracautious in our choice of language when discussing anything about North Korea. *Deferential, respectful,* and *ingratiatory* were the words we lived by. We desperately hoped the North Koreans would not take offense at the secretary's remarks and set us back. But the idea of a release before the June 4 trial date was looking dim.

# CHAPTER SIX

# the phone call

∽∾ **LAURA**

ONE COLD, DRIZZLY AFTERNOON, Mr. Yee and Mr. Baek showed up at my room, and Mr. Yee gave the nod that I knew meant he wanted to go outside for a walk. I put on a black parka that had been provided for me and followed them outdoors. Mr. Yee did not seem himself. His face was flushed and his eyes were glazed. I thought he might be inebriated.

"Is there any news?" I asked.

"I'd tell you if there was any news." His voice had a nasal sound, and he seemed congested.

"Are you okay? Are you feeling sick?" I asked sympathetically.

"Of course I am not okay. Do you think I like being in detention? You are not the only one in detention, you know. As long as you are here, I have to be here, and he has to be here." He pointed to Mr. Baek.

"I'm so sorry. I really wish you could go home, that we could all go home. I know my actions have affected a lot of people, and I feel terrible about that."

"Your government, I just don't understand them. Your black president"—he often referred to President Obama with those words—"he doesn't seem to care about you."

"So what do you think is going to happen?" I asked nervously.

"I've been drinking today," he said. "Do you know why I've been drinking?"

I shook my head.

"I was very upset today because I thought that by now your government would have done something. Every day, my bosses tell me to wrap up this case so they can put you on trial. I've been dragging things along so your government might act. But now it's too late."

"What do you mean it's too late? What's going to happen to me?"

"You will be going to trial."

"Does that mean there's no hope left? Will I be going to a labor camp?"

"Don't worry. It won't be for very long."

"But you said that even the U.S. media is reporting that we'll receive a harsh sentence if we go to trial."

"You will be given a long sentence, but you'll probably only serve for a year or two before your government finally decides to act."

"What? A year or two? I can't be here for that long. I won't last. I won't survive!"

"Don't be silly. You'll be fine. It will be good for you. It will make you tougher. Nelson Mandela was in prison for twenty-seven years."

I was in shock. Tears streamed down my cheeks, and my legs began to buckle underneath me. I crouched to the ground to regain my balance. The rain was pouring down around us. I couldn't fully comprehend what he was telling me. Had the North Korean authorities decided it was too late to work out something with my government?

"Don't be like that. Get up. You have me, your big brother here," he said trying to console me. "Let's go inside."

He escorted me back to my room and turned away to leave. As he did, Mr. Baek translated his words: "I really wish I could communicate with you." I could tell there was more he wanted to say, but he couldn't with Mr. Baek around.

I looked at Mr. Baek and asked him if he knew what was happening.

"Laura, have you heard the saying 'There's truth in wine'? It means the truth comes out when you've been drinking. I think he's telling you the truth."

"It can't be true," I protested. "I can't go to prison. I need to see my family."

"I'm so sorry," he said with downcast eyes. "You really are here at the worst possible time. Relations between our countries haven't been this bad since the Korean War."

I retreated to a corner in my room, curled myself up into a ball on the floor, and wept uncontrollably. A few days later, I was informed that a trial had been set for early June. I was also told that I would be allowed a second visit with Ambassador Foyer, and that I could give him the letters for my family and bosses at Current TV.

When we arrived at the Yanggakdo Hotel, I was led to the same floor where I'd met him before. It was May 15. A month and a half had passed since my first meeting with the ambassador. During that time, I'd been grilled over and over by Mr. Yee and had confessed to attempting to bring down the North Korean government. Now I was set to stand before a judge in court.

Pictures of Kim Il Sung and Kim Jong Il seemed to be watching my every move. There were several small conference rooms, and I was taken inside one of them and told I would wait for about fifteen minutes. I could hear the ambassador's voice outside in the hallway. He was being led into one of the other conference rooms, and I suspected Euna must be meeting with him first. Imagining Euna in the same building as me, perhaps just one or two doors down, made me feel jittery and emotional. I wanted desperately to see her, to run out of the room and hold her.

When Ambassador Foyer finally came into the room, I felt like I was being reunited with a long lost family member, even though I'd met him only once before. I could see he was searching me with his eyes to get a sense of my physical condition. Knowing that this meeting was critical and that I might not get another chance to see him, I repeatedly spoke of the need for urgent action to avert the trial.

"I know that if we go to trial the sentencing will be very harsh," I said, trying to hold back tears and maintain my composure.

The ambassador looked at me reassuringly and said in his soft-spoken voice, "You should understand that a trial is not necessarily a bad thing. It is part of a process."

"But, Ambassador, they are going to send us to prison for a very, very long time, perhaps for most of our lives. Please try to see if my government can do something before the trial."

"Laura, a number isn't always what it means, remember that," he responded. He said he was requesting that he be present at the trial.

I handed him an envelope containing my letters and hugged him tightly before being escorted out of the room.

I T WAS TWO MONTHS to the day since Laura and Euna were captured, the morning of May 15. I was driving to the gym when my BlackBerry buzzed with a message. It was an e-mail from Linda, and it had an attachment. The subject line read: "letter from Laura." I pulled over to the side of the road, parked my car, and then hit DOWNLOAD ATTACHMENT. As it was opening, I could see a scan of a handwritten letter. It was Laura's writing! I was too excited to go to the gym and turned my car around to head home; I had to open the letter on my computer.

Linda's e-mail said Laura had given the letter to Ambassador Foyer. It was my first real contact with my sister in eight weeks. Laura had actually been able to get five letters out through the ambassador: one to Joel Hyatt of Current TV, one to her colleagues, one to Iain, one to our parents, and one to me. The content of the letters varied greatly. To Hyatt, Laura urged the need for diplomatic intervention and asked if he could get Vice President Gore to help. She didn't know that Gore was already involved. To Iain, Laura apologized for making work such a priority. She wrote, "If I'm lucky enough to come home, I promise, no more traveling for me."

To our parents, Laura wrote that she was okay. She said she was being treated fairly and asked them not to worry about her and to take care of themselves. She also wrote eloquently about how she spent her days.

> . . . When I first got here, I cried so much. Now I cry less. I breathe deeply and think about the positive things that happened in the day. For example, I think, "I'm lucky I've gotten through another day," "I'm lucky I'm not in prison," "I'm lucky I saw a butterfly." . . . Each night I am thankful to have gotten through another day. And each

*day when I wake up, I hope it is bringing me closer to home . . . Stay
strong and please take care of yourselves. That is my biggest request.
Know that I am doing okay and dreaming about being reunited with
you all again. . . .*

The letter to me was more urgent.

> *. . . As I'm sure you know, I am in the worst possible situation. I
> have confessed to some very serious crimes that are regarded as hostile
> actions toward the DPRK. And while I have expressed my deepest
> regret for my wrongdoings, I'm not sure it will be enough to send me
> home anytime soon. I am scared.*
>
> *I am desperately hoping that some sort of serious diplomatic in-
> tervention by the U.S. government can be made before our cases go to
> trial. I fear that if something at a higher level is not done soon, I will
> find myself serving a very long prison sentence. That is a thought that
> is too hard to bear.*
>
> *I love you so much, Li. You've looked out for me my entire life. I
> am so fortunate to have such an amazing big sister. I know the whole
> family leans on you for support. Be strong. . . .*

The difference in the letters was a window into my relationship
with Laura. She didn't want to worry our parents, who she knew
were already devastated and petrified. I imagined my sister in a
room alone, filled with stress over the right language to employ
in her first communication to her family. It was painful to think
about Laura nervously but methodically trying to think of the
exact words to use. I could see her penning the letters using her
left hand, and I thought about the days when we were kids and our
grandmother used to try to force her to use her right hand. She
just couldn't. Laura was born a leftie and would always stay that
way, to our grandma's chagrin.

I know she was scared. I could see the fear in her perfectly written letters. This was her only way to reach me, and she needed me to understand the seriousness of her situation. Laura knew that among our family members, I was the one who could possibly move the needle. I had connections in the political world and in media, and Laura knew I would not give up until I got her out of there.

Mom was continuing to leave highly emotional phone messages and send correspondence to Minister Kim at North Korea's Permanent Mission to the United Nations, also known as the "New York channel." To her great surprise, on one occasion Minister Kim actually picked up the phone. He expressed his regret for what was happening, but he told Mom that my sister's situation was not in his hands, rather in those in Pyongyang. She said he seemed genuinely regretful and she made a point of saying that he actually sounded like a kind man. Nevertheless, the next day and every day thereafter, Mom followed up with a letter, fax, e-mail, and call to him.

Meanwhile, the political situation on the Korean Peninsula was growing worse and worse. North Korea was threatening to test a nuclear device for the second time in its history. In response, the global community, led by the United States, warned of severe consequences should it go through with its intended actions.

On May 25, the U.S. Geological Survey reported a 4.7-magnitude quake in the northeastern part of North Korea, around the town of Kilju. Geological agencies in both South Korea and the United States said the tremor indicated a nuclear explosion. North Korea had made good on its declaration to resume its nuclear program and had tested a nuclear device.

President Obama immediately issued a strongly worded statement charging that North Korea's pursuit of nuclear weapons and ballistic missiles threatened peace and was in "blatant defiance of the United Nations Security Council." He went on to say: "The danger posed by North Korea's threatening activities warrants action by the international community. We have been and will continue working with our allies and partners in the six-party talks as well as other members of the UN Security Council in the days ahead."

The United States was not alone in lambasting North Korea for its alleged defiance of the world powers. French officials said they would push for new sanctions. British Prime Minister Gordon Brown called the test a "danger to the world." Russian authorities compared the power of North Korea's bomb to that which obliterated Hiroshima and Nagasaki, and its foreign ministry called the explosion "a serious blow to international efforts to prevent the spread of nuclear weapons." Even Pyongyang's more frequent defender, China, said it was "resolutely opposed" to the test.

The United Nations Security Council immediately convened another emergency session. But this time, few believed that sanctions would not be imminent. North Korea seemed to be defying the whole world, including its allies, by its acts of purported aggression. This was one of the tensest periods ever in U.S.–North Korean relations. Never before had both China and Russia made such public condemnations of their Communist ally. My sister was in the middle of a full-blown nuclear standoff.


∿∾ LAURA

E
VEN AFTER I HAD made the confession the North Koreans wanted, the investigation continued with more questions about our documentary project. Now Mr. Yee wanted to know about the Internet sex worker we'd interviewed at our hotel. I told him what I asked her and how she responded. I also gave him a general description of her appearance. "She's a little shorter than me, with

long black hair. She wore black boots and a white jacket and had on a lot of makeup. I would guess she's probably in her midtwenties." I could have been talking about half the girls in China. But whenever he asked me for the names of sources and interview subjects, I always gave the same answer: "I don't recall her name. I just referred to everyone as 'Sir' or 'Miss.'"

My unwillingness to cooperate about names of sources angered him, but for me it was not an option to put these people in any more jeopardy than they were already in. The only name I volunteered was that of Andrei Lankov, whom we had interviewed in Seoul. Mr. Lankov is a well-known authority on North Korea who has written lengthy papers about it and speaks publicly about his research. I knew they would not be happy that we'd spoken to him, because of some of his critical assessments, but I wanted to give them at least one name, and I knew Lankov would not be in any danger. I wanted to seem as if I was cooperating to the best of my ability.

One day Mr. Yee brought in the pocket-sized tan notebook I'd been carrying during the trip.

"Is this yours?" he asked.

"Yes," I replied.

"Did you use it to write down any notes during this trip?"

I knew I needed to admit to destroying and tampering with evidence when Euna and I were at one of the detention facilities along the border. I told Mr. Yee what we'd done, including swallowing the sheets from my notebook. He flipped through the pages and saw some notes I had jotted down from previous assignments. One page had the name Rory Reid and a list of questions that I had asked Reid, the chairman of the Clark County Commission in Nevada, during an interview I conducted about the declining economy of Las Vegas.

"If you've written his name down, you must have written the names of the people you interviewed in China," he charged.

"Sir, I knew the report about defectors was sensitive in both China and North Korea. That's why I didn't document anyone's names. I didn't want to endanger anyone."

I was telling the truth. The only names I had in my notebook were Pastor Chun's and Andrei Lankov's. The notes I'd destroyed were for the interviews with Chun and Lankov, and they had to do with the regime's stability and whether it would survive for very long. I knew such questions would infuriate the authorities. So while it didn't look good that we had gotten rid of the pages, I felt glad we had.

Part of the investigation required me to submit a written confession of what I had conveyed to Mr. Yee in our daily sessions. The confession had to be written neatly—if I made any mistakes, or crossed out any words because of a misspelling, Mr. Yee would make me rewrite the page. I ended up penning a draft and then copying it again in perfect handwriting. Over the course of a week, I wrote more than one hundred pages, sixty-five of which were part of the final confession document. I woke up in the morning and wrote throughout the day.

One afternoon, as I was hunched over the desk in the guards' room with pen in hand, Min-Jin walked over and handed me a few pieces of candy. "Don't tell the other guard," she said. Touched by this gesture, I slipped the round, sugary ball into my mouth. When Hyung-Yee entered the room, I quickly put the candies in my coat pocket and went back to my writing. Soon after, Min-Jin left the room to take a break, and I was left with Hyung-Yee.

While Hyung-Yee did not speak any English, we still found ways to communicate on a certain level. She was a sweet, lighthearted girl who took pleasure in singing patriotic anthems at the top of her lungs. Sometimes I would catch her mimicking me. If I coughed, she coughed. When I sighed in desperation, she sighed. I'd snap my fingers and she'd follow along. She never continued with these antics for very long, just long enough for us to share in a moment of amusement.

After Min-Jin walked out of the room, Hyung-Yee approached me, opened a drawer in the desk, and pulled out some candy. She handed me some corn candies and flavored rice snacks. She pointed to the couch where Min-Jin usually sat and waved her finger from side to side. I understood immediately that she didn't want me to tell Min-Jin that she had shared the treats with me. "Thank you, thank you," I said to her gratefully in Korean. She popped a sweet into her mouth, smiled, and went back to reading her book.

At the end of each day, Mr. Yee would collect whatever confession pages I had finished and inspect the handwriting. After I had completed the confession to his satisfaction, he showed up with a red ink pad and instructed me to fingerprint each page. I was told that the trial would take place on June 4, just two weeks away.

ONE DAY MR. YEE gave me a brief lesson in North Korean Communist thought and the official ideology known as Juche.

"The nucleus of Juche," he explained, "is that man determines his own destiny. Man is responsible for his actions. He is the master of everything. Do you understand?"

"Yes. So according to Juche, I must determine how I am going to get through this and get home."

The corners of his lips turned up ever so slightly. "That is correct," he replied.

I tried to apply the philosophy of Juche to my own situation. I knew I couldn't just sit by and wait for something to happen. But there was only so much I could do while being confined in a room. More than anything, I wanted to contact Lisa. Even though she was considered an enemy of the state, I knew she could use her connections to have more attention paid to our situation, something the North Koreans might care about more than anything. On a number of occasions, Mr. Yee had mentioned how quiet our families and the U.S. government were being. He said he believed the U.S. government was muzzling our families to make the story disappear from the public consciousness.

He also said that though some people in the North Korean government would send us home under the right circumstances, other hard-liners wanted to send us to a labor camp immediately. If Lisa could get the media to focus on our situation, the North Koreans might see this as movement toward some action.

The problem was that I didn't know how to reach her. I didn't want to send another letter, which was a purely one-way conversation. I decided to ask Mr. Yee if I could telephone Lisa. I didn't think any previous American detainee had been allowed to call home, but during one of our walks, I tried to convince Mr. Yee that allowing me to speak with Lisa might yield results and get them what they wanted.

"You will be able to listen in on our conversation," I said. "We can go over everything I'm going to say beforehand. There are things I can convey to her that I can't say in a letter or to the Swedish ambassador."

"Like what?" he asked.

I mentioned Lisa's relationship with various media figures, including Oprah Winfrey and Jon Klein, the president of CNN.

"If the U.S. government is really trying to keep things quiet, they may not want to hear me telling Lisa to go to the media," I said. "Do you see why it's better to make a phone call?"

I didn't really believe the U.S. government would care what message I had about the media, but I was desperate to hear Lisa's voice and to let her hear mine. It was the most convincing reason I could think of.

"I understand," he replied, nodding. "I will think about it."

Nearly two weeks after my first request to call Lisa, Mr. Yee told me that I might have an opportunity to phone my family. He asked what would be the best time to call. I figured it would be best to call them in the evening, which was around 2:00 P.M. Pyongyang time.

One morning, he came to my room and asked to go over what I wanted to say to Lisa. My main message was for Lisa to try to ignite some media coverage of our case. I wanted to stress the need to act swiftly before a trial, which I'd been told would most likely end in the worst possible way.

Once again I was taken to a conference room at the Yanggakdo Hotel, where a telephone was placed on a lace doily on top of a long wooden table. Before we got to the hotel, Mr. Yee had told me I could call Lisa, Iain, and my parents. But when we arrived at the hotel, he changed his mind and said that there wasn't enough time to call my parents. I pleaded with him until he finally consented. Mr. Baek left the room to listen in on my calls and presumably to translate the conversations for the authorities. I was left alone in the room with Mr. Yee.

Just picking up the telephone receiver seemed slightly strange. I hadn't used or even seen a phone in more than two months. I carefully dialed my mother's number and anxiously waited for her to answer. After a couple of rings, I heard her voice. I was so overcome with emotion that it was hard to speak. But I didn't want to let her or my father hear me cry. I could only imagine the heartache my parents were experiencing from my absence. I was their little girl. I struggled hard to hold back my tears. I didn't speak with either of my parents for more than one or two minutes. But just hearing their voices and allowing them to hear mine was a blessing. I assured them that I was being treated fairly and that I was holding up okay.

When I spoke with Iain, I broke down. His soft voice crackled with emotion.

"I miss you so much," I sobbed.

"I miss you too," he replied, in a voice so weak it was almost hard to make out his words. He asked me how I was being treated and if I was getting his letters.

"I'm okay," I said. "I'm being strong. And, baby, your letters are keeping me going, thank you." I told him that I thought about him every second of the day, and that I reserved that special time at 9:00 A.M. for us to be together.

"That's our time," I said. "No matter what happens." I also asked that he continue to include in his letters as much information as possible about any geopolitical goings-on that might factor into our situation.

"I'm sorry that I risked everything in our life together," I cried.

"Don't you ever give up," he replied. "We are never going to give up."

Then it was time to call Lisa. I was told I'd have roughly ten minutes to speak. I dialed Lisa's number and waited anxiously for her to pick up.

᪷ **LISA**

THE DAY AFTER NORTH KOREA'S nuclear test, Paul and I were tuning in and out of an *Iron Chef* rerun when the call came. It was 10:15 P.M. on a Tuesday, more than two months since Laura was captured. The caller ID on my cell phone said it was coming from a blocked number. I figured it was my mom or dad, the only two people I know who have blocked numbers.

I was a little agitated by the late-hour call, but I answered the phone to hear a distant, little voice.

"Hi, Li, it's me!"

"Oh my god, Baby!" I exclaimed. "Where are you?"

"I'm still here," my sister's voice replied. "I'm still in North Korea. I'm scared," she continued, breaking into tears.

"Are you okay, sweetheart?" I asked. "How are you being treated?"

"I'm okay. I'm being treated fairly." Then she said, "Li, I need your help."

My head was spinning, and I practically froze and couldn't speak. All of a sudden I became certain that our conversation was being monitored and that we had to be careful of what we said. I didn't want to ask anything that would make Laura's situation worse, but I wanted so much to know if my participation in the National Geographic documentary had doomed her in any way. I also didn't know if Laura knew that the country holding her captive had just tested a nuclear weapon. I decided to let Laura do most of the talking.

"Li, I have confessed to some very serious crimes," she said. "My only hope is that there can be some kind of diplomatic action. Our countries have to talk."

It sounded so simple, and if we'd been discussing almost any other country, this "talking" could definitely have been achievable. But North Korea's government had chosen the path of isolation—even from its closest ally, China. It had stopped communicating with everyone. So why had they let Laura call me? Were the country's leaders now talking through Laura? Over and over again she kept telling me that the United States needed to talk directly to North Korea.

"There needs to be some kind of diplomacy," she said several times.

The message they were sending through my sister was clear. I so wanted to tell her that both Al Gore and Bill Richardson had volunteered to go to Pyongyang to get them, but I wasn't sure if I should use names over an open phone line. The State Department had told us that the North Korean government had never allowed calls from other Americans who had been detained. This was part of the game. I figured that it was about to be my turn to communicate with the people holding my sister.

I furiously scribbled notes as Laura repeated the need for a dialogue between the United States and North Korea. It became clear to me that this was not going to be an easy release. By allowing my sister to talk about diplomacy, the North Korean government seemed to be communicating that they wanted something, and whatever it was, it was going to have to be big. I wanted to speak to Laura, but I also had to figure out what to say to the North Korean leadership.

"Baby Girl," I said, trying to say what I thought those listening in wanted to hear, "I am so sorry this happened. We are all so sorry. I really hope your captivity can be kept separate from the geopolitical issues. They are two totally different things and it just wouldn't be right if they were tied together. Have faith, baby. I truly believe in the goodness of people and hope you and Euna will be allowed to come home to your families. That would be such a positive gesture to show the world."

I then told Laura that all of us had kept quiet and had not spoken publicly since her detainment. After a slight pause, and in a soft but deliberate voice, Laura replied, "I think it's been too quiet."

Laura then said she had to get off the phone. Hanging up with her was one of the hardest things I ever had to do. It had been more than two months since I had heard her voice, and I didn't know when I would hear it again. But my orders from my sister were clear—it was time to go public. Laura also asked me to share what she'd just conveyed to me with Iain and my parents. Her captors had let her call them as well but very briefly.

Right away, my parents, Iain, and I got on a conference call to compare notes on our individual conversations with Laura. I recounted everything she had told me and reiterated the point that it was time to break the silence. Laura essentially said the same thing to both of our parents, which was that she was okay and that she missed them very much. Iain didn't share much of what Laura had said to him other than her sorrow that she had ruined their lives.

Every one of us agreed that it was time to go public. This came straight from Laura's mouth, whether they were her own words or those of the people holding her. I phoned Kurt Tong at the State Department and Al Gore to tell them that we were setting out a plan of action. I didn't really give either man a chance to object. I explained that the message from Laura was clear, and our family had unanimously decided to move forward and go public. Gore did interject a disclaimer.

"I personally don't think it's the right thing to do," he said, "but I cannot stop you if that is the decision you and the family have made."

As soon as I got off the phone with the former vice president, I consulted a number of professors of North Korean studies as well as diplomats about the message. The general consensus was that if our message came out wrong, it could inflame the North Korean government. If the regime found itself challenged in any way, it could make matters worse for Laura and Euna. Everyone, except Governor Richardson, advised against talking to the media.

The governor told me he'd learned from his previous dealings with North Korea that its leadership sought out publicity. I wondered if this could be a way for the North Korean government to save face, given the volatility in the political spectrum. Perhaps we could give the regime a reason to make a kind gesture, by publicly apologizing on behalf of the girls and their families and asking for forgiveness. Governor Richardson thought this approach might work.

"If they agree to release the girls, they'll want the act to be acknowledged as an act of humanitarian goodwill," he explained.

It was a hugely risky endeavor, and I think most of the high-level people involved feared we would stick our feet in our mouths. But I couldn't let that happen. On Monday morning we would launch a full-blown media campaign. My twenty years in the television business were being put to the test. This would be the most important story of my life. We had to convey Laura's message, and I knew her captors would be watching.

∿∽ LAURA

I WAS EUPHORIC AT THE SOUND of Lisa's voice. I clutched the phone so tightly during our conversation that my hand hurt. It was as if I were hoping I could break through the line and reach her. I knew she was doing everything in her power to get me home. I knew we would give our own lives for each other.

I thought back to the night before I started kindergarten. We shared a bedroom and were nestled in our beds. One of our nightly rituals involved us imagining ourselves in a spaceship traveling to far-off lands and galaxies. We had a robot named Irona, just like the one in the Richie Rich cartoons. Together, Irona, Lisa, and I would ward off evildoers and save the world from destruction. In one fantasy adventure, Lisa, who usually did the narrating, started giving Irona and me our instructions. "Now it's time to say the Pledge of Allegiance!" she ordered. I was silent. I didn't know the Pledge of Allegiance but was too embarrassed to admit it.

"Lau!" Lisa shouted. "It's your turn to say the pledge!"

"What's the pledge?" I asked.

"Oh my god, you don't know the Pledge of Allegiance? What are you going to do tomorrow at school?"

Suddenly our spaceship ritual didn't seem so much fun. I didn't understand how this pledge could affect my first day of school. Lisa told me I needed to be prepared and began going over the lines of the pledge until I had memorized each word. "See, it's not so hard," she said. "You're going to be great." We were each other's protectors, coconspirators, and best friends.

One thing during our phone call still confused me. "Lau, things are really complicated," Lisa had said. "Do you know about the nuclear test?"

"Yes, I know. I know it's a tough time right now. Just do what you can," I replied.

The words "nuclear test" didn't register with me while we were on the phone. I thought Lisa was referring to the satellite launch I'd seen on the North Korean news weeks earlier. But later, as I was running the call through my mind, I thought, *Did she actually say "nuclear test"?* I thought back to the various news reports I'd seen on television. There was always some rally or another going on. I felt certain that one of them had to be in celebration of a recent nuclear test.

⁙ **LISA**

I HAD TO ACT FAST. My sister's trial was coming up in nine days, so we had to do everything right. Our family had gotten requests for interviews from every news outlet imaginable, but we had turned down all of them. Now we had to be strategic. My friend Alanna Zahn, a longtime publicist, took on our effort full-time and for free. She contacted the *Today* show because it was the biggest morning news show in America and one of the hosts, Meredith Vieira, was a colleague of mine from our days as cohosts of *The View*. It was important that we be interviewed by people who understood how careful we had to be.

The *Today* show agreed to bring our family on to make our plea. I asked Euna's husband, Michael Saldate, if he would join us with their four-year-old daughter, Hana. I could only imagine how hard it would be for Michael to put his daughter on TV and through a media maelstrom, but we needed to have little Hana be a part of our media blitz. If the girls' captors were watching, I wanted them to see this little child in anguish over the absence of her mother. Euna had not conveyed the same information to Michael that Laura gave to me, so he was skeptical about going public. Michael didn't tell me too much about his conversation with Euna because it seemed to be much more personal in nature. However, he finally relented about going to the press after much persuasion from Iain and me. We were grateful; we needed Hana to make our appeal.

There would have to be a CNN play because I learned from former chief news executive Eason Jordan, who had visited North Korea years ago, that the Dear Leader watches the network and receives a direct feed. Wendy Walker, executive producer of Larry King's show, told me that we could have the hour to make our appeal.

In our message, Iain and I were very careful not to blame the North Koreans. We had to be careful not to accuse them of having done anything wrong. We wanted to suggest that this might be a unique opportunity for North Korea and the United States— two countries without diplomatic ties—to communicate directly. Most important, we needed to stress that our issue should be kept separate from the larger geopolitical and nuclear issues. We also made a conscious decision to refer to Laura and Euna as "the girls" to somehow convey to the North Korean leadership a sense of vulnerability about them. We didn't want the North Koreans to use Laura and Euna as leverage for their nuclear ambitions. As we were preparing to leave for New York, we combed through our notes, and I gave my parents specific and emotional talking points that we rehearsed over and over.

On the plane to New York, Iain and I went over and over the points: "We are sorry. When they left the United States, they never had any intention of crossing the border into North Korea. Only Laura and Euna know what happened on the border that day. This could be a unique opportunity for direct dialogue and diplomacy. We miss Laura terribly. We are very concerned about Laura's recurring ulcers."

The morning of our *Today* show appearance, Meredith's flight got delayed from where she was reporting, so Matt Lauer ended up interviewing us. I had known Matt for a long time, and before the interview started, I let him know about the sensitivities involved. We were dealing with such an unpredictable regime that we could not afford to deviate at all from our meticulously crafted message of apology.

The show went off without a hitch. My parents were terribly nervous and insecure about speaking publicly, and Matt was understanding of that. He directed the political questions to Iain and me, and together we were able to make our plea for compassion to the North Korean government. He asked my parents about how much they missed their youngest daughter and talked to Michael about how hard it was for Hana to be without her mother.

Larry King was next. We all jumped back on a plane to Los Angeles to tape the CNN show live. The hour-long program was to break down with me in the first block of the show, Iain and Euna's husband joining me in the second block, and my parents and me in the last. In my haste, I had forgotten to ask the producers to please tell Larry not to ask about my experience in North Korea.

During my one-on-one with Larry, he recalled that some years ago I had been in North Korea.

"You've been there before," he remarked. "What was it like?"

My heart started pulsating rapidly. "Uh, yes, I was there years ago and . . . uh, the people were very kind."

"What was it like?" Larry pressed.

"Well, it was interesting and the people were kind," I repeated. "It's not really germane to what's going on."

All of a sudden, Larry looked distracted. I could tell the control room was speaking to him through his earpiece; someone was telling him to stop asking about this. He stuttered a bit before quickly changing the topic.

Aside from that moment, the show went swimmingly. Michael didn't want Hana to be on the set with us, because hearing about her mom's detainment earlier that day on the *Today* show had upset her. So we had a producer stay with her in the greenroom with toys. From time to time throughout the broadcast, the show cut to the greenroom to take shots of Hana playing with her toys. We stayed on message and made our emotional, nonpolitical pleas. We had officially gone public.

After our interview with Larry King, I went into the greenroom to remove my makeup. I saw that there was a message on my BlackBerry. It was from Kurt Tong from the State Department.

"Nice job," the e-mail read. "The tone was perfect."

I didn't hear from Al Gore at all the next day, so I decided I would phone him to get his thoughts. He picked up right away.

"Was it okay?" I asked nervously.

"Lisa," he said, "you all did a magnificent job."

**D**URING ONE OF OUR next walks, I asked Mr. Yee if he'd heard any news from the United States.

"There's some news," he replied as we continued to stroll the walled compound.

I was caught off guard. Weeks had gone by without there being any encouraging information. I asked nervously, "What kind of news, good news?"

He said my family had appeared on various news programs in the United States, including *Larry King Live* and the *Today* show. My sister had come through with her promise to go public with our story.

"That's good, isn't it?" I asked.

"We shall see," he answered. "Your family is working hard, but it's your government that needs to act.

"Your sister," he added, "she's really brilliant."

His demeanor had changed, and I saw this as an opportunity. I desperately hoped the North Koreans would see what could be gained by letting me speak with Lisa again. I was determined to figure out what it would take to get us released and how I could convey this to her in a future call.

"Was there a nuclear test conducted recently?" I asked.

With one eyebrow raised and a smirk, Mr. Yee responded, "How did you know about that?"

"My sister mentioned it on our call. At first I thought she was referring to the satellite launch, but after thinking about it, I specifically remember her saying 'nuclear test.'"

"Yes, we conducted a nuclear test. Would you like us to do another one just for you?"

I rolled my eyes at him. "Thanks, but no thanks," I responded dejectedly. My optimism was starting to fade away. Did the North

Korean authorities seriously think the United States would make any concessions for us after this defiant act?

Mr. Yee noticed that I had become sullen. "What is the matter?" he asked.

"It just keeps getting worse," I replied. "First a satellite launch, now a nuclear test. We're caught in the middle of this political game. I wish our situation could be separated from the politics, but I know that's not possible."

"You're right," he responded. "Our countries are still at war, so we can't just release you. But your government will do something eventually. There's no use in keeping you here forever."

"What about Roxana Saberi?" I said, referring to the Iranian-American journalist who had received an eight-year sentence. "She got eight years in prison. And for what, allegedly drinking some wine? Our case is much more serious than that."

"Roxana was released from prison," he said.

I couldn't believe what I was hearing. Her release had not been reported on the North Korean news. Was it really true?

"What?" I asked eagerly. "What happened?"

He explained that President Obama had made an overture to the Iranian government and that Roxana was released on appeal after about two weeks. "She's home now," he said.

I struggled to contain the smile forming on my lips, but it was no use. I was grinning from cheek to cheek.

"Really?" I said. "That's amazing. That's the best news I've heard in a while."

I could see a slight smile on Mr. Yee's face, but then he went right back to his resigned expression.

"Time to go inside now," he said.

After that, days went by with no information.

The phrases "Stay strong," "We'll see you soon," and "Everyone is doing all they can" seemed like standard signatures on every letter. Just days before the trial, I received a letter from Lisa dated May 19, 2009, in which she described various candlelight vigils that would take place in different cities across the United States on the day of the trial.

> *Hi Baby Girl,*
>
> *I believe I wrote to you about this guy Brendan Creamer on Facebook who has taken up your cause and has orga-nized vigils all over the country to show support for you and Euna . . . It's pretty amazing . . . It has been inspiring to see how much support you have around the country and the world.*

From the letters, I had already read about some vigils that had taken place in different cities, including one at my high school in Sacramento and one organized by a college friend in Los Angeles. It was strange to think that so many people were lending their support. While I'd never been more isolated in my life, thinking about all of the prayers for Euna and me made me feel a little less scared.

IN THE DAYS LEADING UP to the trial, Mr. Yee asked if I wanted to have a defense attorney appointed to my case. I accepted, mainly because I was curious to see what the attorney could do at such a late date, after I'd already been interrogated for two months and had issued a sixty-five-page written confession.

It didn't take me long to figure out that the attorney was really just an extension of the prosecution. His main questions had to do with my treatment in captivity. "Have any of your human rights been violated here?" he asked. "Have you been tortured?" He never asked anything about the actual case. I could tell from his questions that I shouldn't open up to him. Instead, I just let the process take its course.

After I met with the attorney, Mr. Yee asked me what I thought of the man.

"He's fine," I said dispassionately. "Although it seemed odd that he spent only fifteen minutes talking with me and never once mentioned the actual case or my charges."

Two days later the attorney returned and spent an hour asking me questions about the investigation. He wrote down notes and feigned interest in what I was telling him. His visit this second time was a reaction to my comment to Mr. Yee. It was clearly important to the North Koreans that their legal process appear as legitimate and genuine as possible.

As the trial date neared, Mr. Baek reminded me, in an excited, upbeat tone, "You are the first U.S. citizen to be tried in a North Korean court. You are making history!"

"Thanks," I responded. "But I'd rather not be making history this way."

I asked Mr. Baek if he would be translating for me at the trial. He said he hadn't been told that he would, and he believed his assignment was supposed to end before the trial. This sent me into a panic. I hadn't imagined anyone other than Mr. Baek at the proceedings. I desperately wanted him there, not only because I trusted that he would interpret everything accurately and efficiently, but also because his presence put me at ease. With him by my side, I wouldn't feel so alone.

I asked Mr. Yee about this and told him it was absolutely necessary that I be comfortable with whoever was translating, that the slightest misinterpretation might skew the case and make the trial seem flawed.

"I understand," Mr. Yee responded immediately. "I will explain your situation to the appropriate people."

Finally I was told that my request had been granted. Relieved, I said to Mr. Baek, "Looks like you'll be making history with me."

Although I dreaded the trial, I was happy that I would finally get to see Euna after being separated for two and a half months. But I was also told that Ambassador Foyer would not be allowed to witness the proceedings, nor would Mr. Yee.

⁌❀⁍ **LISA**

**B**RENDAN CREAMER, WHO SET UP the Facebook page Free Laura and Euna, arranged ten vigils in different cities across the United States to coincide with the June 4 start of Laura and Euna's trial. In addition to our many friends, throngs of people whom we'd never met organized in their respective communities to offer their support. Across the country were Meghan in Portland, Lisa in Montgomery, Rose in Chicago, Paula in Phoenix, Danielle in New York, and Clothilde in Paris, among so many others. There would be a fair amount of press in attendance, so we asked Brendan to tell the organizers that the tone of the vigils had to be respectful; this was imperative. Iain, my parents, and I decided to attend the gathering in Los Angeles, along with Michael Saldate. We wanted to be wherever people were out supporting the girls.

I had started my journalism career with Anderson Cooper in the early 1990s at Channel One News. Now Anderson had a CNN prime-time show called 360 that was simulcast live around the world. It even aired, coincidentally, at the top of the morning in Pyongyang, and although the average North Korean citizen is not able to access CNN, we hoped the officials there would see it. I asked Anderson's producers if the show would broadcast from the vigil in Los Angeles, where our families would be in attendance.

It was 7:00 P.M. Pacific standard time in L.A. on the night of June 3 and 12:00 P.M. in Pyongyang on June 4. With hundreds of people standing behind us solemnly holding candles, Iain, Michael, and I made our respectful plea for leniency. From his anchor desk in New York, Anderson stated that "as we speak," the trial of Laura and Euna had likely begun.

"We believe that CNN is seen in North Korea," Anderson reported. "Lisa, is there anything you'd like to say to the North Korean government if they are watching?"

My insides were trembling. This was my chance to appeal to the humanity of my sister's captors. I took a deep breath and recited the scrupulously crafted script once again. "We can say with absolute certainty that when Laura and Euna left U.S. soil they never intended to cross the border into North Korea. If at any point they committed a transgression, then we profusely apologize on their behalf. We know they are terribly sorry. We beg the government of North Korea to show mercy and allow the girls to return home to their families that miss them desperately." I concluded by saying, "And perhaps this can be an opportunity for our two countries to engage in more direct diplomacy."

We couldn't know if the North Korean leadership would hear our plea, but if they did, we had to make sure we hit the right tone. I didn't want them to see these vigils as protests, or rallies—we didn't want to make angry demands. Our greatest hope was for Laura and Euna's trial to conclude with their captors deciding to let them go.

## ∞ LAURA

ON THE DAY OF THE TRIAL, I nervously paced back and forth in my room, my palms soaked with sweat. I filled my pockets with tissues that had been sent to me by my family, and I even packed extra for Euna. I changed into a new short-sleeved, button-down blouse that had been given to me for the proceedings. It was pink with thin white stripes. I also wore brown slacks and sneakers that had been provided. My eyes were bloodshot because I had been up all night anticipating every question they could ask me and rehearsing conciliatory statements I could give in return. I knew there wasn't going to be any presumption of innocence. In the eyes of the North Korean government, I was already deemed guilty of trying to bring down the regime. My strategy therefore was to agree with them and be as respectful and apologetic as possible. By telling them what they wanted to hear, I was hoping they might show leniency.

The dog let out a series of loud barks as a vehicle pulled up into the compound. I watched as two female soldiers dressed in green fitted army uniforms entered the room and sternly instructed me in Korean to look down to the ground. My heart began to palpitate when one of them pulled out a pair of handcuffs. As she wrapped the cold metal cuffs around my wrists, I thought back to the day of our capture along the frozen river when Euna and I were handcuffed and transported to jail. Now I was again being enveloped by that same feeling of entering the dark unknown. I felt I was edging closer to a deep abyss. As they led me out of the room, I shifted my gaze to my guards.

Over the past two months, we had developed a mutual respect for one another and even a sort of kinship. Confined in a room together for twenty-four hours every day, we became used to one another's habits and personalities. One of them always sang at the top of her lungs anytime she was in the bathroom, and the other had a tendency to burp loudly. I came to care about them, and I believe they felt the same way about me. As I left the room in handcuffs, Kyung-Hee looked at me sympathetically, while Min-Jin turned away and closed her eyes. It sounded as if she might have been crying.

I was transported a few miles away in a small van. One soldier sat beside me on the middle seat, and one was in the back. Once again I was ordered to close my eyes and hold my head down. I meditated for the duration of the ride. We arrived at an office building, and I was taken up to the second floor into a large room. The area contained long rows of chairs situated before a stage. Three men sat at a table on the stage. On the floor below were two tables, one for the prosecutor and one for the defense attorney. Along with these key figures there were two videographers, a photographer, my interpreter, the soldiers, and a couple of authorities milling about in the back of the room. The room wasn't like a traditional courtroom; it looked more like a performance space that had been set up for the purpose of putting us on trial.

As I was led to the front of the room, I caught sight of Euna's small frame. It was incredible to see her, and to know she was alive. Her head was hanging low in a defeated posture. Her hair had grown out and was now long enough to be tied back. We were wearing identical outfits. I was instructed to sit next to her and to continue looking downward. I couldn't tell if Euna even realized it was me sitting beside her. I scooted close enough so that my thigh touched hers. I wanted to grab her hand and hold it, to look her in the eyes, but I was too nervous to do anything but sit there in silence.

Euna was instructed to approach the podium in front of us, where she was asked some basic questions in Korean. Then it was my turn. Aided by Mr. Baek, the judge confirmed my name, profession, and the charges against me—illegally entering the country and committing "hostile acts" against the Korean nation. After acknowledging and accepting his statements, I was led out of the room. Euna was to be tried first, and I would follow.

After a couple of hours, I was escorted back inside. I was disheartened when I saw that Euna was no longer there, but at least I had gotten a quick look at her.

I approached the podium, and the prosecutor, a towering figure with wavy hair and a perpetual snarl, began grilling me.

"Do you know who Agnes Smedley is?" he began.

"No, sir," I replied, wondering where his questioning was headed.

"You don't know who Agnes Smedley is?" he asked again with a smirk. "She's an American."

"I'm sorry, sir. I don't know who she is."

"It doesn't surprise me that you are unfamiliar with Agnes Smedley, because Agnes Smedley was a highly respected journalist. She wrote about the Chinese civil war in the 1930s. She exposed the truth, and the Chinese even built a statue to honor her. I'm not surprised that you don't know who Agnes Smedley is, because she represents what a good journalist should be like." He went on about the heroic deeds of Smedley and her positive portrayal of the Chinese Communist revolution. "You call yourself a journalist, but you do not know how to do your job like Agnes Smedley."

I could see that the prosecutor was relishing this opportunity to humiliate me, and I suppose he wanted to throw me off guard. But it didn't bother me. I wasn't prepared to put up a fight. Throughout my interrogation, I had refined the art of obsequiousness. I knew that showing respect to those in authority was of the utmost importance.

"You're right, sir," I responded courteously. "I am not a good journalist, and I am very sorry for my actions."

The prosecutor continued to ask questions that had already been posed to me during the investigation, including the purpose of my story and what motivated me to work on the documentary. As rehearsed with Mr. Yee, I took a deep breath and told them I wanted to help the people of North Korea, even if it meant bringing down the North Korean regime, which I believed was oppressing its people. It was hard to believe that my work on a documentary about human rights abuses had taken me to North Korea's highest court, where I was confessing to having tried to topple the government.

During the investigation, Mr. Yee had gone to great lengths to denounce defectors as criminals with bad intentions. He had explained that these "scum of the earth," as he called them, were creating lies about the DPRK in order to make money and hurt the regime. To prepare me for the trial, Mr. Yee wanted to make sure I understood and believed that the defectors were unscrupulous.

"I understand," I replied dishonestly, "the defectors we interviewed were lying. They were criminals in the first place in North Korea, which is why they might be put in prison if they are caught."

"Precisely!" he shouted.

I felt I was being indoctrinated into the North Korean belief system, being fed the propaganda that the government disseminates across the nation in an attempt to maintain a unified society. I told him what he wanted to hear—and kept telling him until he was satisfied.

During the trial, I repeated these denouncements of North Korean defectors. Every time I did, I thought about Ji-Yong, the twenty-something girl who had escaped from her country only to be lured into the Internet sex industry. I'll never forget the look in her eyes as she told me about fleeing a life of horror and devastation. These defectors were people I had hoped to shine a light on for their bravery and courage, and now I had to malign them verbally. Each time I said something bad about them, I felt sick.

When they asked how I felt about Pastor Chun Ki-Won, the missionary who had aided us on our project, I castigated him by saying he was using the defectors to raise money for his own gain. As painful as it was, I knew I had to condemn Chun and the entire defector network in order to prove to the court that I understood my wrongdoings and the implications of my actions.

Even though Mr. Yee had prepped me for the prosecutor's questioning, it was torturous to be in the room with the stern-faced judge peering down at me, knowing that my fate would be decided in a matter of days. All of this put me on edge. From my dealings and interactions with North Koreans for the past few months, I had personally seen how regimented they are. They seem to conduct themselves publicly in a very formal, almost robotic way. I wanted to maintain my composure in the courtroom, so as not to interrupt the disciplined mood, but I couldn't help breaking down. I tried to hold back my tears, but it was no use. When I lifted my hands to wipe my cheeks, my fingers knocked into the microphone in front of me, causing the speakers to screech. I shifted from one foot to the other in agitation and was ordered by one of the soldiers to stand still. I could barely speak. The bellicose prosecutor seemed to derive great pleasure out of making me squirm. Meanwhile, my defense attorney sat silent.

Mr. Baek was normally calm and confident, but today he was noticeably nervous. This was, as he had explained on so many occasions, a historic moment, and he didn't want to make any mistakes in his interpretation. He scribbled down notes feverishly. His voice cracked at times, and large beads of sweat dripped from his face onto his notebook. I looked at him with deep appreciation. He'd never done anything like this before, and I wondered what the consequences might be for him if he made a blunder.

After two hours, the prosecutor, smug and haughty, concluded his first day of questioning. I exhaled in relief, grateful to be released from his verbal blows. I never thought I would look forward to going back to the compound, but the growls of the guard dog, which usually gave me shivers, now sounded comforting because the trial was over—at least until the next day.

The proceedings would last two more days for a total of about eight hours. On the final day, a large screen was set up to show the video evidence they had obtained from our belongings. The first part shown was footage from Euna's video camera on the brisk morning of March 17 when we were filming on the frozen river. It was eerie to look at the images, which showed life before we were apprehended. I looked like a timid cat as I walked along the ice following our guide to a place and a moment that would change our lives.

For a brief instant, I found myself looking at the images as a reporter, and I was able to admire the scenes Euna had captured. *This could have been a really eye-opening documentary,* I thought. As I watched the footage, I imagined the risks so many thousands of people are taking to escape the desperate conditions in their homeland. I was disappointed that I wasn't able to bring that story to light.

"Did you cross the river onto DPRK soil?" the prosecutor asked forcefully.

"Yes, sir. It was only a few steps, but I did trespass, and I am very regretful for my actions."

Then it was the defense attorney's turn to question me. He began with an odd inquiry: "What did your parents say to you before your trip?" I answered that my mom and dad told me to be careful, as they did before I left on any assignment. His questioning perplexed me. I didn't see how the conversation with my parents had anything to do with my case. He followed up with other irrelevant questions. I looked at him in disbelief. I had to keep myself from bursting out laughing at his absurd line of inquiry—this was my attorney, the man who was supposed to be standing up for me! Finally, as if to fulfill his role as my defense lawyer, he asked if I was sorry for my crimes.

As the trial drew to a close, the prosecutor and defense attorney each had an opportunity to give their closing statements. Rather than talking about my specific situation, the prosecutor began by speaking about U.S. imperialism and the U.S. government's constant meddling in North Korean affairs. My actions, he explained, were an extension of the U.S. government's policies toward the DPRK, which should be viewed as a threat to the North Korean regime. He argued for the strictest sentencing of fifteen years.

Though I had been expecting the worst, I held on to a sliver of hope that the trial might result in a full release. The words "fifteen years" echoed inside my head. All of the mental exercises I had done to help prepare me for this moment didn't do me any good. I was petrified. I remembered my shock at hearing about Roxana Saberi's eight-year sentence in Iran. My legs wobbled, and I grasped the podium for support.

The defense attorney began his closing statement by chastising me for crossing the Tumen River. He said that if I had applied for a visa, I would have been allowed into his country legally. The only thing of value he did was acknowledge that I seemed genuinely sorry for my actions and argue that I should be given a more lenient sentence.

I was then told to be seated while the judge and his associates left the room to discuss the sentencing. I tried to remain calm as I waited for the judge to return with his verdict. I closed my eyes and visualized the candlelight from the vigils taking place back home. I conjured Ambassador Foyer's words, "A number isn't always what it seems." He had also said that the trial was part of a process toward our eventual release.

After a mere five minutes of deliberation, the judge returned and took his seat on the stage. I was told to stand up and walk to the center of the room, where he could face me head-on. My heart was thumping fiercely. The judge began reading from some sheets of paper. In a thunderous voice, he explained that I was being given a combined sentence of fifteen years in a labor camp. He then went on to say that the sentence was being reduced to twelve years, which included two for trespassing and ten for "hostile acts." He never mentioned when we'd be transferred to the camp.

I felt my world closing in on me fast and furiously. Everything around me started spinning.

"There will be no forgiveness and no appeal!" the judge exclaimed.

I fought hard to hold back a flood of tears. It wasn't the sentence's harshness that shocked me. I had predicted all along that they might give me a long prison term to send a message to the outside world. It was the phrases "no forgiveness" and "no appeal" that tore into me. Did this mean they were closing the case, even if the U.S. government offered some sort of gesture?

I was led out of the room while Euna completed her last portion of the trial. After about an hour, I was brought back in front of the judge with Euna. The judge went through various formalities such as restating our crimes and the twelve-year sentence. We were instructed to sign and fingerprint documents verifying our acceptance of the outcome. I knew that the end of the procedure was upon us, and it wouldn't be long before Euna and I would be separated. Fearing I might not get a chance to see her again, I embraced her tightly. We were both sobbing hysterically. I envisioned Euna's little daughter, Hana, who would be without a mother. I thought of my own parents who would soon find out the news about their own daughter. The soldiers split us up and escorted us out of the room separately.

# CHAPTER SEVEN

# the window is closing

◦§§◦ LISA

L AURA AND EUNA'S TRIAL DATE, Thursday, June 4, came
and went with no news out of North Korea. We didn't even
know if the proceedings had begun or not. For more than
two and a half months we had been anticipating the trial date,
and now it felt almost anticlimactic because we had no information. Still, I was a complete wreck with worry and could hardly
get my mind on anything else. On Sunday, I was scheduled to
deliver a commencement address at National University in San
Diego that had been scheduled for more than a year. In some
ways, it was good to have something to do. I didn't say anything about my sister's situation while onstage, even though I
knew people wanted to know the latest. A couple of news crews
showed up to see if they could get a comment, but I simply told
them I had nothing to say.

That night, some friends insisted that Paul and I go with them to a movie as a way of getting my mind off what was happening. They chose *The Hangover*, thinking it would provide the highest level of distraction. Only twenty minutes into the filmic drunkfest, the red light on my BlackBerry started flashing. Someone I know at ABC News was sending me an urgent e-mail.

The subject: American journalists sentenced to twelve years hard labor in North Korea. I immediately forwarded the e-mail to Iain before I was overtaken by a sense of panic.

I suppose I was naive, but somehow I believed that our apologies and passionate appeals for mercy would convince Laura's captors to pardon her and Euna and release them after the trial. But a labor prison? I showed the message to Paul, and we got up immediately and left the theater.

I called Al Gore right away and told him the news; he was at his home in Tennessee.

"Damn them," he said. He seemed to be as shocked as I was.

The phone call with Gore was brief; there was an eerie silence as we tried to figure out what to do next. We agreed to call each other as soon as either of us got any more information, and then we hung up.

I called Kurt Tong on his cell phone, and he was just hearing the news. In an hour it was everywhere.

Iain, Paul, and I headed to my mom's house. By the time we all arrived, it was nearly 11:00 P.M. Kurt set up a midnight conference call with our family, Michael Saldate, Linda McFadyen, and a number of other State Department people. It was 3:00 A.M. in Washington, D.C.

My mom was in hysterics and crying loudly into the phone. With his always carefully worded State-Department-speak, Kurt led the call.

"Look, we are all surprised by the severity of the sentence," he said, "but we don't want anyone to panic. We believe that this will give us an opening to communicate in real terms—which is a good thing. The North Koreans are trying to show that they have a legitimate legal process and have reached completion in the trial of the women."

"How do you know?" my mom pressed. "What if they send her to a labor camp, what will we do?"

"We hope that does not happen," Kurt replied. "Hopefully dialogue will now begin."

Kurt's urging us not to overreact didn't work. We were all devastated, especially my mom. We hoped Kurt was right, that the verdict was a sign that talks could happen soon between the United States and North Korea.

In the days after the sentence was announced, the news was filled with stories of what my sister might be forced to endure in one of North Korea's brutal gulag-style prisons; no American had ever been sent to one. I went online and found a disturbing report about North Korea's labor prisons entitled "The Hidden Gulag: Exposing North Korea's Prison Camps," compiled by David Hawk for the U.S. Committee for Human Rights in North Korea. It provided details of life inside labor camps across North Korea, where an estimated two hundred thousand people are alleged to be living:

> The most salient feature of day-to-day prison-labor camp life is the combination of below-subsistence food rations and extremely hard labor. Prisoners are provided only enough food to be kept perpetually on the verge of starvation. And prisoners are compelled by their hunger to eat, if they can get away with it, the food of the labor-camp farm animals, plants, grasses, bark, rats, snakes—anything remotely edible.

Now my sister, who just a couple of months before had told me that she and Iain were working on starting a family, had been sentenced to serve twelve years in one of these camps. I wondered what her captors had told her. I was gravely concerned about her mental state; I could feel my baby sister's fear and anguish from a world away.

SPIRALED INTO A DEEP depression. I refused my meals and rarely moved from a chair in a dark corner of the room. I envisioned spending twelve long years in a North Korean labor camp. I worried about Iain not having a partner in his life. I tried to imagine women I knew who might make a suitable match for him. It crushed me to think of him being with someone else, but I didn't want him to be alone. I wanted the best for him. I thought of my father's weak heart and feared I might not ever see my parents again. *At least I'll have Lisa when I get out,* I thought. I imagined myself, weathered and gray by the time I was released, moving into Lisa's house and helping her take care of her children. That would make me happy, I smiled to myself. While I tried to sketch out my life of imprisonment, thinking of how I might endure the long dark days of isolation, I also contemplated suicide. *I will try to get through at least two years,* I thought.

Min-Jin approached me with a tray of food. "You need to eat," she said.

"I can't," I responded, tears falling from my cheeks. "I'm not hungry."

"Just try a little something," she persisted.

"Do you know what happened? Do you know what sentence I was given?" I asked.

She placed the tray down and sat on the seat beside me. "No, I don't know," she said curiously.

"Twelve years."

Her eyes widened in disbelief. "Twelve years?" she said softly.

"Yes, and they said, 'no forgiveness and no appeal.' I can't survive twelve years in prison. I need to see my family."

She was speechless. I could tell she was genuinely shocked by the news and saddened for me.

"I thought you would be forgiven," she said compassionately. "I thought you would get to go home."

I sensed that she was starting to feel uncomfortable sitting beside me, consoling me. It wasn't her place. After all, she was supposed to be guarding me. But I could tell she felt disappointed. I think Min-Jin really believed her government would act compassionately toward me. I could see she felt awkward just leaving me there alone.

"Thanks for listening to me. I really appreciate it. I'll be okay," I told her.

After sitting with me for a few more minutes, she got up and went back into the guards' area, where she began whispering to Kyung-Hee about my sentencing. I could hear Kyung-Hee's surprised reaction, followed by a somber silence that washed over the room.

Later that day, Mr. Yee came to take me for a walk outside. As he waited outside while I gathered my coat, Min-Jin called out to me.

"Laura," she said softly, "have hope."

I was incredibly touched that these words were coming from the woman who was supposed to keep me prisoner.

Outside, as we began walking along one side of the compound, Mr. Yee asked me about the proceedings.

"It was terrible," I said. "Twelve years."

"Did that surprise you?" he replied. "I told you that even your own media has been reporting that you would get a very long sentence."

"I suppose I knew it might happen, but I was secretly hoping for forgiveness."

He chuckled softly. "And how did you find your defense attorney?"

"What defense attorney!" I said cynically. "He might as well have been working for the prosecutor! But it's as I expected. I know he was just doing his job."

"That's right, he was. We don't have attorneys like you do in the United States, where they get paid a lot of money. This man did his best, and he got your sentence reduced to twelve years from fifteen. I saw him afterward, and he said he was sorry he couldn't do more."

I felt bad about insulting the attorney who I knew was powerless in a system that was using Euna and me as pawns for larger political purposes.

I asked Mr. Yee about the conditions at the camp. "You will be fine," he responded. "You are a journalist, after all. Now you will get to see what a real prison looks like in North Korea." I had become used to his snide remarks.

"Is the camp in Pyongyang or close to Pyongyang? I really hope we're not moved too far away from the capital," I said.

"I'm not sure exactly where you are being sent, but I think you are going to the prison that's about an hour's drive away."

"Do you know when this will happen?"

"That I don't know. There are still some procedural things that need to be done."

Mr. Yee then questioned why I hadn't been eating.

"I can't eat. I'm too upset," I explained.

"You must eat something."

"How can I eat after a verdict like that? The judge said, 'no forgiveness and no appeal.' What does that mean for us?"

He shrugged off my question.

"In Roxana's case in Iran, there was an appeal," I continued.

"That's how she was able to return home. Does 'no forgiveness and no appeal' really mean there is no hope left?"

After taking several steps along the compound wall in silence, Mr. Yee finally said something that brought me back to life. "Law does not determine things. Man does," he said, looking me in the eyes. "Do you understand?"

I did. Suddenly I began to see things through a different lens. Even though Euna and I had been subjected to a trial and given a sentence that was most certainly preordained, the ultimate power lay in the hands of the regime. I wondered if the "man" Mr. Yee was talking about was Kim Jong Il or someone else who was jockeying for power. I wanted to figure out what needed to happen for the person who was calling the shots to overrule the law. Our fate was still murky, but I was relieved to know that, as Min-Jin had said, there might still be room for hope.

⋘ **LISA**

EVERY DAY SEEMED to present one surreal thing after another. I went to bed every night hoping that the next morning would bring positive developments in my sister's case, but each morning I woke up fearing that something terrible might have happened overnight. June 12 presented a doozy.

Sending a clear and decisive message, all fifteen members of the United Nations Security Council voted unusually but unanimously to impose the severest sanctions yet on North Korea. Susan Rice, America's ambassador to the UN, called the sanctions "unprecedented," and said the United States was very pleased with the vote. The resolution strengthened an arms embargo and called on the international community to search North Korean ships for weapons and materials that could be used to boost its nuclear and ballistic missile programs.

Not surprisingly, Pyongyang responded by threatening to declare war on any country that dared to stop its ships under these new sanctions.

Days later, Secretary of Defense Robert Gates stated during a press conference that North Korea's threats were being taken so seriously that U.S. defense forces had been ordered to prepare for a possible missile launch toward Hawaii.

Right after Laura and Euna were captured, if someone had told me that U.S.–North Korean relations would get this bad so quickly, I would have said it wasn't possible. This seemed like a cruel game, and all I could think was that Laura was one of the pieces. But in reality, it was anything but a game. We were quite literally in the midst of a global nuclear showdown, and every day seemed to bring worse news. Tensions were growing on all sides. I decided then that I needed to find other avenues to get Laura out, even if it meant trying to infiltrate North Korea.

Around this time, I had a conversation with a friend who told me that someone she knew was trying to get in touch with me about my sister's situation. All she would say about the man was that he was part of an organization affiliated with the U.S. government that performed highly classified special missions.

I got the phone number of the man I'll refer to as Brad and called him. Brad claimed that he had been a part of JSOC, the Joint Special Operations Command. JSOC consists of some of the most elite covert and clandestine operatives from multiple agencies and branches of the military in the United States. Brad told me that for years he had been working to infiltrate terrorist organizations in Afghanistan, Iraq, and South America. He used the phrase "human intelligence" a number of times. He thought he could help me. He said, "The phone is a dangerous device to those in my line of work. We live and work underground."

We didn't discuss particulars because of security concerns. But he did let me know that sometime in the course of six weeks, his team, which consisted of other ex-JSOC members, would be able to "get eyes" on my sister. In other words, his team would determine exactly where Laura was being held. He would put this information together and present it to American intelligence agencies and military organizations. They would then try to determine what course of covert action was required to get Laura and Euna out.

But there was no guarantee. It would cost five hundred to fifteen hundred dollars per day, depending on the expenses incurred. "There is huge risk to this," Brad said repeatedly.

When I asked how the operation would be conducted, Brad told me there was an "extensive underground cellular ring" in South Korea that pushes "throwaway" phones into the North. The phones are used to communicate between people separated by the demarcation line. He said his team had already made contact with some in this underground network who had been operating in it for years. They were ready to take action and find out where Laura and Euna were being kept. They just needed the "go-ahead." I figured that meant having a financial transaction take place.

That's all he would tell me by phone. He warned me that failure to protect his identity would lead to major problems for him and provoke alarm within what he called "the system." He said the greatest risk to him was that he would be "blackballed" in the network in which he operated. I swore I would never mention his name to anyone, ever.

I told Brad I wasn't sure the risk to my sister's safety was worth it. What if Brad's network got compromised? Could his efforts possibly jeopardize Laura's situation and make matters worse? I wanted to give diplomacy a little more time, even though it hadn't done much good in the past few months. It seemed as if every time we thought we had an opening, the door got slammed in our faces. So I wasn't about to rule out the Brad option, or any others for that matter.

~◦~ **LAURA**

**W**ITHIN A FEW DAYS after the trial, I began to feel an intense, burning pain in my stomach area. I had also been vomiting uncontrollably whenever I ate something. I feared my ulcer was flaring up and asked Mr. Yee if I could see a doctor.

"You will be going to a hospital for a full medical checkup," he replied. "It's part of the process before going to prison, to see if you are fit to serve in the labor camp."

The next day I was taken to a hospital in Pyongyang. Mr. Yee remarked that I was becoming a true North Korean because I was going to experience North Korea's fine medical treatment. The hospital was clean but sparsely furnished. The dim hallways were empty, and I didn't see any other patients around. I was introduced to one of the medical directors, a short, middle-aged woman doctor with a gentle manner. She asked me about my ulcer and examined my stomach area. I winced as she lightly pressed into my upper abdomen.

Speaking through Mr. Baek, she explained that I needed to undergo an endoscopic procedure to look into my stomach and check for ulcers. This was familiar to me because I had gone through several stomach "scopes" in the United States. The process involved having a doctor place a thin tube down my throat into my stomach area. At the end of the tube was a small lens that sent images back to a computer monitor. I was given anesthesia during these procedures in the United States, so I never felt a thing.

I looked around the room at the limited and outdated equipment, and suddenly became frightened that this procedure might not go the same way as the ones I'd had in the United States.

"I'm a little nervous," I told the director. "I'm not so sure it's a good idea to do the scope. Will I be given anesthesia to put me under?"

"It will feel just like it did in the United States," she assured me.

"Well, back home, I didn't feel a thing because of the anesthesia. Can you make sure to give me some?" I said nervously.

Mr. Baek, sensing my jitters, chimed in, "You really are a baby girl, aren't you!"

He was using the affectionate nickname Lisa called me in her letters, which told me he had read them. This didn't really bother me. Mr. Baek always had such a kind, jovial way about him. I was even happy that he knew so much about my family and friends. It made me feel more connected to him.

"Yes, I am a big baby," I said sheepishly.

A nurse arrived and escorted me, along with Mr. Baek, to an operating room where I was given a shot, which I was told was an anesthetic. But rather than numbing my senses, it just made me feel dizzy. I watched as a doctor prepared the instrument that would be placed inside me. It looked more like a hose than the narrow tubing I remembered from past procedures.

A plastic apparatus was placed in my mouth to keep it lodged open. I braced myself as I saw the doctor coming toward me with the thick black tubing, which he inserted down my throat. The feeling of the instrument making its way down into my body was so agonizing it sent me writhing. Several nurses rushed to my side and pinned me down. I continued to gag in pain, unable to scream because of the plastic device. I began coughing and releasing large gusts of air through my throat.

One by one, several doctors took turns looking down the hose into my stomach and consulting with one another while I continued to struggle on the table. Periodically they shifted the hose to look at different sections of my belly. Drops of sweat fell from my forehead. I clenched my fists together, closed my eyes, and thought of Iain. *Please help me get through this, baby,* I thought. Finally the doctors began pulling the hose back up and through my mouth. I sighed deeply in relief.

After the procedure, I was led into the medical director's office, where she was sitting with the prosecutor from the trial. I cringed at the sight of him but forced myself to bow toward him respectfully. He nodded with a slight snarl and motioned for me to take a seat. According to the medical director, my ulcer was quite serious, and I had developed several stomach lesions. She recommended that before I was sent to the labor camp, I be given a few weeks under a doctor's supervision.

"She doesn't look like she's in any pain!" the prosecutor yelled. "My wife has an ulcer and can function just fine."

In a calm, even-toned voice, the medical director refuted the prosecutor's assessment and made her opinion clear. She did not think I should be sent to prison just yet. I looked at her with deep gratitude.

"You can leave now!" the prosecutor said gruffly, waving his hands in the air. I followed Mr. Baek out of the room, where Mr. Yee was waiting to take me back to the compound.

Afterward, Mr. Yee told me I would remain at the compound and was being placed in medical detention until I was deemed fit to go to the labor camp. I wondered if this was part of a plan to allow more time for my government to act.

"We are not giving you any special treatment," he said, as if reading my mind. "This is part of our legal process. Everyone sentenced to prison needs to go through a full medical checkup to see if he or she is capable of performing labor. Your medical detention is only temporary. It might last a week or longer."

"What about Euna?" I asked, worried that the doctors might not have found anything to prevent her from being sent to prison.

"She is very weak," Mr. Yee explained. "She has some arthritis. I don't think she can go to prison. But they are still deliberating on her case. Your situation has been decided. I don't know about hers."

I WOKE UP EARLY ON June 16 to news that the Korean Central News Agency (KCNA) had released details of the charges against Laura and Euna. The report indicated that both of them had admitted to engaging in criminal activity.

The timing was interesting, given that South Korea's president, Lee Myung-bak, happened to be visiting the United States that day and would soon be meeting with President Obama at the White House. It seemed like a deliberate attempt to upstage the highly publicized meeting. High on their list of priorities was to discuss how to deal with North Korea's nuclear ambitions.

The list of charges delivered on this day would surely be distracting and would remind the United States that North Korea had two of its own. Part of the KCNA statement read, "At the trial, the accused admitted that what they did were criminal acts, prompted by a political motive to isolate and stifle the socialist system of North Korea, by faking moving images aimed at falsifying its human rights performance and hurling slanders and calumnies at it."

I lay in bed wondering what this all meant. I e-mailed Kurt Tong and Al Gore and asked them to try to ask President Obama to treat our issue delicately if asked about it during the press conference on the White House lawn with President Lee. Gore responded that he would try to get the right language into the president's talking points by phoning people he knew in Obama's inner circle. I just didn't want a repeat of the time Secretary Clinton called the charges against Laura and Euna "baseless," publicly calling into question North Korea's legal system.

At the press conference, when a question was asked about the American journalists inside North Korea, President Lee responded by calling on the North Korean government to release the girls along with a South Korean Hyundai worker also being held. Mom, Iain, and I had practically affixed ourselves to the screen to hear our president. We knew his every utterance regarding North Korea was being watched and scrutinized by those holding my sister. Words mean everything. In calculatedly stern remarks, President Obama left the door open for North Korea to come back to the negotiating table.

"There is another path available to North Korea, a path that leads to peace and economic opportunity for North Korea, including full membership in the community of nations," President Obama said. We could only hope the North Koreans would agree to come through the door.

Not long after the sentence was delivered, rumors had started swirling about Al Gore being sent as an intermediary to bring Laura and Euna home. The general consensus was that he would be a great choice to engage the North Koreans. But other than one interview with CNN's John Roberts, in which he said he would do whatever it took to bring the girls home, Gore never spoke about the situation publicly. He maintained that North Korea had to be dealt with sensitively, and whatever diplomatic efforts took place had to be conducted privately. The former vice president seemed to be making progress. During our Friday conference calls with Kurt and the State Department team, we were led to believe that Gore's people were speaking and possibly even meeting with representatives of North Korea; it was something.

Meanwhile, it had been some time since I'd spoken with Governor Richardson. I had gone from talking to him once, sometimes twice, a day to speaking with him every week if that. After a while, he knew something was up. It was the beginning of the third week of June when I got a call from the governor.

"I think I'm being edged out," Governor Richardson called me to say. "I think they're going with Al. Have you heard anything?"

"Uh, I really haven't heard anything specific," I responded.

I wasn't lying, but I wasn't being entirely truthful with him. I felt sick that I was straddling both camps, but I didn't feel like I had a choice; my sister's future was at stake. Though I know he would have helped regardless of whether or not he had been asked officially, this probably would have restored Governor Richardson to the national and world stages. I truly wished he could have had that chance. But it was becoming clearer and clearer that his role in our situation had been usurped.

Meanwhile, on the Gore front, no one—including the former vice president—could confirm anything with absolute authority. But there at least seemed to be communication. Kurt told us that State Department sources indicated that a Gore visit was being considered, but nothing was definitive yet. We didn't know whom the "sources" were talking to; we were just told that we would have to wait. Waiting. It's all we seemed to be doing.

Though our mom was an emotional wreck throughout the ordeal, at least we were all around to keep her company. During this time, Paul suggested I fly up to Sacramento to spend a couple of days with my dad. He had made several trips down south to be with all of us at our mom's house, but he had his own house to maintain and appointments to attend to back home. I noticed the toll that Laura's absence had taken on our dad when I went for a visit. While at his house, I wrote this in a letter to Laura about him:

> . . . I'm sitting on Dad's black leather couch in Sac. He has aged a lot in the last few months. He tries to be his usual silly self, but there is an emptiness about him that is undeniable. He sits and just stares out of the window for long periods of time without saying a word. When you return, we have to make more of an effort to come see him; it means the world to him. . . .

I HAD BEEN PERSISTENTLY BEGGING to call my family, and later in June I was told that in a few days I could make another set of phone calls to them. I was overjoyed. As before, Mr. Yee asked what I planned to talk about. I told him I wanted to update them on my medical condition and tell them they had to act quickly before we were sent to a labor camp.

"Your family will want to know if you have been moved to another location," he said. "All you need to tell them is that you are now in medical detention and being treated fine."

I was escorted back to the Yanggakdo Hotel with Mr. Yee and Mr. Baek. Aside from being transported to and from the trial, I had only been taken outside the compound grounds on three occasions—twice to meet with Ambassador Foyer and once to call my family. Each time the car exited the iron gate, the guard dog jumped and yelped hysterically. The hotel couldn't have been more than a few miles from the compound because the drive seemed to last only about five or ten minutes.

Inside the hotel conference room, Mr. Yee told me I could speak to Lisa and Iain for approximately ten to fifteen minutes, but I had to make my calls to my mother and father briefer. I dialed my mother's number and began to choke up the second I heard a ringing tone.

I knew I didn't want to tell my parents anything about my medical condition. I was more concerned about their health than mine. When my mom asked about my ulcer, I simply replied, "I'm doing fine. Don't worry about me, just take care of yourself."

"I'm going to make your favorite watercress soup when you come home. And I know it will happen soon," my mom said.

When I called my father at his home in Sacramento, his voice started to quiver once he realized it was me on the line.

"You have to be strong. Your dad loves you," he said, referring to himself in the third person as he often does.

Hearing my mother's and father's anguish was too much to bear. I hung up the phone and started wailing.

When it was time to call Lisa and Iain, I knew I had to get it together emotionally. I saw my conversations with them not only as a rare chance to express my love, but also as an invaluable opportunity to send and receive messages that could be crucial to getting me home.

I'm sure I was given more time with Lisa and Iain because speaking with them was one of the only ways the North Koreans could let the U.S. government know what they wanted in return for our release. I tried to give them both roughly the same information as a way of making sure nothing was lost.

As Mr. Yee had suspected, one of the first questions they each asked was "Have you been moved?" I tried to deflect the question, while assuring them I was okay. I took a risk in telling Iain, "Don't worry, baby, I still look at the sky every morning at our nine A.M. meeting time and think of you." I hoped this subtle hint would not make the authorities mad.

-ᐧᐧᐧ LISA

O N JUNE 21, MY PHONE RANG. It had been two weeks since the severest sentence imaginable was handed down to Laura and Euna, and we had not heard a word about how they were doing since. The press was speculating that they might have been moved. One theory suggested that they had been sent to a prison in Pyongsong, one of the three cities I had visited during my trip to North Korea. My family was gravely concerned about Laura's state of mind, given the harshness of the sentence. We were traumatized ourselves and couldn't even imagine how Laura was feeling after learning that she might not see her family again for more than a decade.

I probably shouldn't have asked Laura where she was, but I had to know.

"Where are you?" I probed. "Have you been moved?"

I put my phone on speaker so Paul could listen in. I was incredibly emotional, and we wanted to make sure I didn't miss or misinterpret anything Laura said. He kicked me to stop me from asking questions that might compromise her.

After a brief pause and in an obvious attempt to avoid answering my question, Laura replied, "The conditions here are decent."

Paul was right. I had asked Laura a question she could not answer. I hoped I hadn't gotten her into trouble, but her captors had to understand how concerned we were. I was in awe of my sister's composure during what I'm sure was a heavily monitored conversation. I wondered how she was maintaining her strength after being told that she was sentenced to a labor prison; my heart hurt so much at the thought.

Laura said she had been given a mandatory medical examination to determine if she was fit to go to prison. An endoscopy performed there in North Korea revealed that she did have an ulcer, and she would be given a bit more time for her health to improve before she was sent away.

The year before, I had accompanied my sister to have an endoscopic procedure in Los Angeles. A tiny scope was inserted down her throat to check inside her stomach. She was put under heavy anesthesia for the procedure, but even so she felt terribly ill afterward. Given my experience in North Korea with the Nepalese medical delegation years before, I shuddered at the thought of the medical facility and antiquated treatment that my sister likely had to endure.

"I feel okay," she said, "but the prosecutor was there with me at the examination, and he kept telling the medical staff that I'm well enough to go to prison. I'm really terrified."

To reassure her and to send a message to those listening in on the call, I told Laura I believed that discussions were being held at the highest level about her situation.

"Our side is ready to do what it has to do," I urged, "including sending someone over immediately to come get you."

In closing, Laura said one final thing. "The window is closing," she said. "We need to move from the talking stage to the acting stage."

As soon as I hung up the phone, I called Al Gore. I told him about the endoscopy and particularly that Laura had sent a message with her comment about the window closing and the need for action.

"They're trying to drum up angst from the family," he said.

He was always extremely careful about what he said on the phone, and I wondered why he sounded so certain now. Then he said something that excited me because it seemed as if definitive contact had been made.

"We may be close to a break," he said. "We have to just sit tight." But before what he was saying could sound too positive, he added, "But we can't be sure, because there may be more than one institutional player in Pyongyang."

This alarmed me, because I was pretty sure he was referring to certain suspicions I'd heard about the leadership in North Korea— in particular about Kim Jong Il and who was calling the shots there. Reports were coming from South Korea that the Dear Leader was very ill, possibly on his deathbed, with pancreatic cancer.

Kim Jong Il hadn't been seen publicly in many months. There was also scuttlebutt in the press about an alleged power struggle between hard-liners and more moderate voices inside the Foreign Ministry. To add to the confusion, rumors were circulating that Kim Jong Il's youngest son might be taking the helm of the country. Some reports indicated that the twenty-five-year-old was more of an extremist than his father, while others speculated the opposite.

Theories and hypotheses about who was in control were being spewed from every direction. But one thing was certain: no one knew for sure.

Around the time I was hearing reports that Kim Jong Il might be sick, I got a call from Michele Chan, who is married to Dr. Patrick Soon-Shiong. I had met Michele and Dr. Soon-Shiong at the annual fund-raiser for Saint John's Hospital in Santa Monica, where my husband, Paul, is a radiation oncologist. Dr. Soon-Shiong is a billionaire, and he recently donated tens of millions of dollars to rebuild the entire hospital. He was being honored the night I met them. According to his company biography, "Dr. Soon-Shiong developed the first FDA approved protein nanoparticle delivery technology for the treatment of metastatic breast cancer."

Michele told me that the drug was also being developed for lung, melanoma, gastric, and pancreatic cancer. She said her husband's company, Abraxis BioScience, wanted to offer the revolutionary treatment to Kim Jong Il—entirely free of charge—in exchange for the safe release of Laura and Euna. I was incredibly touched by Michele's call and offer. I got in touch with Kurt Tong at the State Department right away and asked if this was something to consider.

"It's an interesting thought," he said, "but bringing up the treatment to the North Koreans would be asking them to acknowledge that the Dear Leader was ill."

He had a good point. Given the formidable face that North Korea likes to project to its own citizens and to the world, the mere intimation that their leader was unhealthy could undermine his dictatorial position. So the idea about exchanging the Abraxis pancreatic cancer treatment for the girls' release was never presented.

## ∿∿ LAURA

A FEW DAYS AFTER the phone calls, I was allowed another visit with Ambassador Foyer. As on the previous visits, I cherished each second with the gentle ambassador. He had already received news from my family about my condition and told me that medicine was being sent from them, which he would pass along to the Foreign Ministry without delay. I didn't know it at the time, but this meeting with Ambassador Foyer would be my last.

Three long months had passed since our apprehension along the border. It was now late June, and the summer's heat was blistering. During a walk with Mr. Yee, he told me that he was no longer going to be in charge of me now that the investigation was over. He said a new person would be responsible for me and that Mr. Baek was also going to be released.

I was deeply affected by this news. It had taken me a while to build up a certain trust and rapport with Mr. Yee. On top of that, I had come to see him as a source of valuable information. I relied on the news he provided about what was happening in my country and how the authorities in North Korea were deliberating.

"I don't want you guys to go," I said pleadingly. "Who will come take me for walks and talk to me, and who will interpret for me?"

"You will be allowed to walk still, and there will be someone who speaks English."

"Will you come and visit me?"

"I can't come every day. My job is over. Perhaps I can come once a week."

"But please remember, you have to keep your promise," I said, referring to our late-night conversation over beer. "I did my part and cooperated in the investigation. You have to do your part and help me go home."

"You will go home eventually. I don't know when, but it will happen. And I will see you when you leave."

"Are we talking months or years?" I asked desperately.

"That is up to your government," he replied.

That evening at dinner, I was given a bowl of Pyongyang's famous cold noodles instead of the usual fare. Looking on while I slurped the chewy strings of buckwheat, Min-Jin asked, "Do you like the noodles?"

"Yes, they are very good," I replied. Indeed, the noodles, mixed with cucumber, egg, kimchi, and ginger, were exceptionally tasty. It was a nice treat.

"They were brought in from a hotel," she said proudly.

I remembered when Mr. Yee had boasted about Pyongyang's famous cold noodles. "One day, I will get you some," he told me once when we were driving to meet the ambassador. I knew he had ordered these noodles for me. As I sipped the ice-cold, savory broth, it dawned on me that this was a parting gesture, that I would not be seeing him again, at least for some time. Even though he had been tasked with interrogating me, part of me felt a real sadness about his absence.

The next morning, while I was looking out the window, Min-Jin approached me with a melancholy expression. "Laura, today is our last day. We are leaving," she said dolefully.

"What!" I exclaimed. "That's terrible! When did you find this out? Do you know what is going to happen to me?"

"I just found out this morning and was told to gather my belongings. There will be new guards assigned to you."

"But I don't want new guards!" I cried. "And how will I communicate with them?"

"There will be someone who speaks English, probably better than me."

"No. Your English is great."

"Thank you for teaching me," she said.

"Where will you go? Will you go back to your old job?" She had never told me what she did professionally.

"Yes," she replied glumly. "But I would prefer to be here." I got the feeling that, in some small way, I had opened her eyes to another world.

As I sat in despair, worried about the new people who would be taking charge of me, Kyung-Hee walked in. At the sight of me sobbing, she too began to tear up.

"I just want to thank you both for being so kind to me," I told them appreciatively. "No matter what happens to me, I will never forget you." They nodded in silence.

"I hope the new guards are not mean to me," I said.

"Just do what they say," replied Min-Jin. "We'll tell them that you are a 'scaredy-cat,' who is afraid of bugs," she said, smiling. I had taught the guards the expression *scaredy-cat* when one of them commented on how much I despised the insects in my room. I'd become an expert at killing the dozens of mosquitoes that invaded my quarters each night.

"Well, if I ever get to go home, I hope that one day we will see each other again under different circumstances," I said.

"I would like that," she replied. "Perhaps we will meet each other abroad."

I got up and hugged her. She seemed taken aback by the gesture and stood upright as I squeezed her.

I was sad to see these women go. We had spent almost every moment over the last three months together in these confined quarters. I remembered how suspicious they were of me on that first frigid March morning, and how intimidated I was by them. Now we were parting as friends. I was going to miss them.

As they were leaving, a small group of people entered the room. I immediately recognized one of the men. It was the judge from the trial. He was accompanied by another man and three women. The judge started speaking, and one of the girls, who was pretty and petite, began to translate. Her English was very good. It wasn't perfect like Mr. Baek's—whose translation was not only flawless but who was able to adopt the moods and tones of the person speaking—but I didn't have any trouble understanding her.

"Do you know who I am?" the judge asked.

"Yes. You are the judge from the trial," I responded.

"Do you know why I am here?"

"No, sir."

"I have received the results from your medical checkup. It says you are not ready to be sent to the labor camp because of your ulcer and stomach lesions. You will remain here and be placed under medical detention. Your sentence will be suspended until you are deemed able to go to prison. Do you have any questions?"

"Yes, sir," I replied. "What is going to happen to my friend, to Euna?"

"She is none of your business," he responded sternly.

"Sir, I don't know if she is fit to go to prison, but I don't want her to have to go to the prison alone. If she is being sent to the labor camp, then I would like to go with her."

My comment seemed to surprise him, and he paused for several seconds.

"Just worry about yourself!" he finally replied sternly.

A series of documents written in Korean were placed in front of me, along with a pad of red ink. I was told they outlined my new status, and I was ordered to sign and fingerprint each page.

The judge pointed to the other man in the room. "This man is in charge of you. He is your guarantor. And these are your guards. If you need anything, you should ask her." He pointed to the woman translating. "She will inform the guarantor. You should not ask the guarantor directly."

I was left with the new translator and the two guards, who

ordered me to go back to my room. I felt as if I was experiencing déjà vu and starting all over again with these unknown figures staring at me contemptuously. I began to walk circles in my room, and they watched every movement I made.

I was now under the jurisdiction of North Korea's highest court. The investigation was over, so there was no real need for me to speak with anyone. I missed the walks with Mr. Yee and Mr. Baek, particularly those toward the end of the investigation when we began to engage in small talk. I enjoyed sharing details of my life with them and was intrigued to hear Mr. Yee's perspective on the United States.

I also found it impossible to create any bonds with the new women in charge. They were different from the previous guards; they seemed more regimented and less educated. I went for days without uttering a word to anyone.

If I was going to break through to any of them, I figured it would be the self-assured translator. I could tell from her dainty skirts and blouses and her elegant features that, unlike the two other women, she was part of a special class in North Korea. I was surprised when I saw that she had a personal cell phone. This was the first time in my months in captivity that I had spotted anyone with a mobile device.

The North Korean government began allowing cell-phone use again in late 2008 after shutting it down in 2004. The regime gave no reason for the ban, but it's believed they wanted to prevent people from disseminating information about the country and its food shortages. Now, cell phones are reserved for the privileged, who must receive permission from the government to own one. The cost of purchasing a phone is out of reach for the average citizen, and the phones can only call within the country.

Chinese cell phones, which can make calls outside of North Korea, do get smuggled into the country, however, and some North

Koreans are taking the risk of using these phones to call relatives who have defected and are now living in China or South Korea. It's believed that those caught using a cell phone without permission could face execution.

When I first heard the high-pitched ring on the translator's phone, I thought I had been transported into another time. I had gotten used to not hearing the constant buzzing and ringing that has become such an ingrained part of our modern society. Her phone seemed like an alien object in a gadget-free world.

This personal phone was not the only thing that told me about her family's unique status. She had an MP3 player that she used to listen to language instruction—she was learning how to speak Czech. Special facial creams, shampoos, mirrors, and makeup overflowed her toiletry basket. Each morning she made instant coffee with cream that she stored in a small refrigerator that had been brought into the guards' area. She also had mango-flavored powder to make juice and, to my surprise, a few bottles of Coca-Cola, a symbol of U.S. capitalism. I nicknamed her Paris after the young Hilton heiress. She wasn't flashy or extravagant, but she was clearly privileged in a society of have-nots. Watching Paris with all her fancy possessions reminded me of the anger expressed by the North Korean defector I had met the evening before our arrest. He had talked about the growing disparity between the elite and everyone else in North Korean society.

Perhaps because of her upper-class status and everything the Dear Leader had provided for her and her family, Paris was compelled to pay special homage to Kim Jong Il. My previous guards spoke of their leader with immense pride, but Paris seemed to especially revere him. She turned up the television volume anytime he came on the screen, which was often. While I was given a single grubby cloth to clean the rooms every morning, Paris reserved a special unsoiled white towel to scrub the portraits of the father and son leaders. And she was meticulous about making sure the pictures hung perfectly.

·⚜· **LISA**

I WAS CONSTANTLY THINKING of ways to get Laura and Euna released. I heard that Kim Jong Il is a big lover of Hollywood movies and that his film collection is one of the largest in the world. This made me think the Dear Leader might be impressed by some of America's celebrity names. I've always been skeptical of celebrity involvement in causes because I've met too many "stars" who use social activism to boost their careers. But ours was a case where Hollywood stars might have more influence than politicians.

One of the performers at the Los Angeles vigil, a talented local musician named David Kater, wrote a song that was meant to rally people behind the movement to free Laura and Euna. It was a soulful piece called "Stand Together" that he composed and donated to us. It gave me an idea.

If I could get someone famous to sing the song, maybe a number of A-list film actors would participate in a music video that encouraged diplomacy and peace, and could defuse tensions between North Korea and the rest of the world. The actors would hold up signs that said things like LOVE, PEACE, DIALOGUE, etc. Months had gone by and nothing was working, so I decided this was worth a try. I phoned the Los Angeles branch of the agency that represents me, William Morris Endeavor, to ask if any of its clients would appear in the video.

"It would only take a few minutes of their time," I said to my agent. "All they have to do is hold up a sign."

I got positive answers from Catherine Zeta-Jones, Forest

Whitaker, Keanu Reeves, and a number of others. Angelina Jolie told me by e-mail that it wasn't something she would typically do, but she would consider being part of the video. She was very gracious and said she and Brad Pitt wanted to do what they could to help. Things were coming together. Now I had to find a singer.

Paul and I were home the night of June 24, thinking about who the best candidate would be. I wanted it to be someone recognized internationally. A few names came to mind: John Legend, John Mayer, Sheryl Crow. Then Paul suggested Michael Jackson. He was exactly the right person, but would he do it? That night, Paul downloaded "Man in the Mirror" from iTunes, and we both sat silently listening to the lyrics and Michael's unique cadence as he sang it. I knew that our longtime family friend Gotham Chopra was very close to Michael. I planned on calling Gotham the next day and asking him if he could reach out to his friend.

On June 25, before I had time to call him, the news reported that Michael Jackson had died of a drug overdose. When I spoke to Gotham a few days later, he told me that Michael and he had spoken about Laura during their last conversation, just weeks before his death. Michael had seen the news reports of her capture and knew Gotham was a close friend of our family.

Late one night, Michael phoned him to ask if there was anything he could do to help. Gotham would describe the last time he spoke with Michael in a piece he wrote for the Huffington Post Web site. It was called "Michael Jackson and Kim Jong Il."

In it, Gotham wrote:

> He [Michael] asked me whether I had had any contact with Laura. I told him I had written her a few letters and had been assured they were getting through. Outside of that, her own family had only heard from her twice—brief monitored phone calls—in the over three months they had been imprisoned. When I told him that, Michael paused. "Do

*you think," he said hesitantly, "that the leader of North Korea could*
*be a fan of mine?"*

According to Gotham, Michael wondered if Kim Jong Il knew of his music. He told Gotham that if the North Korean leader did like him, and if it would help, he would go to the Communist state to perform for him. That chance would never come.

I HADN'T DONE ANYTHING work-related in more than three months, but toward the end of June, I decided to take on a couple of assignments. It was nearly impossible to report stories when the most important one to me was happening in my own family.

Although I am an exclusive correspondent for *The Oprah Win-frey Show*, they can't feature my segments every day, so they allow me to do some reporting for a couple of other media outlets. Before Laura's capture, I had started working on another National Geographic documentary about, of all things, the colossal spike in kidnappings in Phoenix, Arizona, as a result of the violence spilling over from Mexico. A big investigation was going down with the Phoenix Police Department, and my producer asked if I'd like to be part of it. I agreed, knowing that if anything were to happen with Laura's situation, I could be on a plane within hours.

I arrived in Phoenix from Los Angeles at 8:00 A.M., and the investigation was already under way. Hours earlier, a young man in his twenties named Paco had been pulled out of his truck, beaten up, and abducted. In his vehicle was Paco's terrified four-year-old son, who saw everything. After securing information from witnesses in the area, the investigations team reconvened at the police station, and Paco's family and girlfriend were called in. Deeply distressed and forlorn, they were all put into a room to wait to see if Paco's kidnappers would call to ask for a ransom. I was told that this is normal procedure: when someone is abducted, often the kidnappers call the family members within twenty-four to forty-eight hours to demand money for the release. If a call came, the Phoenix Police Department would be ready for it.

Hours went by with no word. The cops let me take Paco's girlfriend into another room to see if she would agree to be interviewed. Her name was Sandra, and she was a wreck. Her eyes were puffy red, and black eyeliner was smudged all over her face from crying and wiping away the tears. She was wearing what appeared to be slippers. She must have left her house immediately upon learning of Paco's abduction and hadn't had a chance to change clothes.

As I watched her sob in anguish, I felt like I was looking at myself. I knew exactly what she was feeling. For months I had been asking similar questions about my little sister. *Where in the hell was she? What happened? Will I ever see her again?*

I started to cry. Sandra looked up at me in confusion. I told her I was the one whose sister was being held in North Korea—she knew the story right away. And then I embraced her. I could feel her body shaking as I held her. We stood crying and holding each other for what seemed like ten minutes. Our circumstances were so different but similar at the same time. The biggest difference, though, was that I knew that Laura was alive. I couldn't say the same for Paco.

The two of us came out after about thirty minutes and walked into a room filled with police officers. This time, we both had black, smeared eyeliner all over our faces. My producer asked me if Sandra had agreed to talk on camera. I looked at him and in a quiet voice replied, "No."

A week later, I took an assignment from ABC's *Nightline*, for which I am a regular contributor. A charter school organization had recently taken over Locke High School, the reputed "toughest school in L.A.," located smack-dab in the middle of the warring gang territories of South Central Los Angeles. A riot had broken out on the Locke campus the previous summer that involved six hundred students. The Green Dot Public Schools charter program came in the next year and imposed a dress code and stabilized the security situation.

All the residents in the surrounding neighborhoods, including the many children, had witnessed or experienced some kind of violence while living in the area. My crew and I shot video and interviews on campus and then ventured out to film around the neighborhoods that envelop Locke High School. Gangs mark their territory by spraying graffiti that covers all the walls, storefronts, and food trucks in the vicinity. These were places that cops would advise people not to visit at night, especially not alone. Some years ago, South Central Los Angeles was considered the homicide capital of the United States. Even in the middle of the day, we were seeing drug deals in progress on multiple street corners, and prostitutes were combing the boulevards for johns.

As I was interviewing a high school student on his street, a strung-out, scantily clad woman who was clearly "working" stopped what she was doing and began to stare at me. I felt a bit uneasy about where I was, and my heart started beating faster. She continued to glare, and then all of a sudden at the top of her lungs she screamed, "Lisa Ling, is that you? Where's your sister, girl?"

There I was in the middle of the "hood," and this woman was asking about Laura. I was moved to speechlessness. I smiled and yelled back, "We still don't know."

"I'm praying for her, girl!" she replied.

Then, no less than five minutes later, a tall, very thin man who looked homeless appeared and began to walk toward me. He looked me up and down before saying, "Hey, have you found your sister? I've been praying for her."

I was in a part of Los Angeles known for violence and mayhem; these were streets where killings occurred regularly, but on that day it felt like home to me. These people, whose lives could not be more different from mine, were saying prayers for my sister despite their obviously challenging predicaments. The reactions I was getting from people from so many different walks of life told me that our story was one that people seemed to rally around. I deeply appreciated the support—and I needed it.

# CHAPTER EIGHT

# glimmers of hope

## ⤳⤶ LAURA

OFTEN, WHEN PARIS WASN'T in the room, one of the other guards would go through her belongings, curiously inspecting objects that were foreign to her. She would apply Paris's mascara to her own lashes, flip through her Czech language book, and dab Paris's foundation onto her skin.

One day Paris left her cell phone in the room unattended. Captivated by the small black gizmo, the wide-eyed guard picked it up gingerly, pressed her stubby fingers onto the buttons, and unintentionally turned off the phone. Confused, she shook the gadget up and down, trying to wake it up. I could see she was getting nervous; Paris could walk into the room at any moment and see her handling the prized possession. Suddenly the panicky guard turned to me, approached, and held the phone out. I was apprehensive about taking it from her in case Paris entered the room and thought I was trying to call someone. But the guard pressed the phone into my hand with a look of desperation. I grabbed the object, wishing I could indeed dial my sister's number and hear her sweet voice. I pressed the green telephone icon, holding the button down long enough for the phone to come on, and quickly returned it to the guard, who rushed back to her area and placed the phone back where it belonged. Within moments, Paris returned. The guard innocently turned the pages of her tattered book, pretending to be engrossed in its story.

Like the two guards, Paris was cold and curt with me when I first met her. But rather than staring at me endlessly as the two guards did, Paris seemed uninterested in my existence. My presence seemed more of a nuisance to her than anything else. She was brought in to interpret for me, not necessarily to guard me, and her uppity attitude conveyed her higher sense of being.

One day I asked if she knew why I was being held in medical detention.

"Of course I know. Everyone in the country knows. It was all over the news," she replied.

I wondered how the government news readers had characterized me and my crimes. "So, what do the people say about my colleague and me? How do they feel about us?" I asked.

"It's not like people talk about your situation in the streets, but they are aware of what you did, and no one is happy about it."

I was hesitant to talk about our documentary for fear that I would be accused of trying to brainwash her. Instead, I said only, "When I was arrested, I was working on a documentary about people who leave North Korea. I wanted to help these people and bring greater awareness to the situation of those who are suffering. I know people here believe I had hostile intentions. I just want you to know that I am sorry if my actions could have hurt anybody."

She listened intently. "The rest of the world thinks that North Korea is a horrible place," she said passionately. "And you probably won't believe me when I tell you that we really like living here. We are very proud of our country and what we have been able to achieve."

I couldn't help but think she was speaking for the elite group like herself who have cell phones and MP3 players.

"We don't like living without electricity and water," she went on. "We know we are not a rich country. But it is the United States that has put sanctions on us and has deprived us of these things. What did we ever do to the United States?" Paris wasn't the only person I'd spoken with who blamed the U.S. sanctions for North Korea's lack of electricity. This was a common theme discussed by my other guards and Mr. Yee. To them, every blackout—and they happen multiple times a day—reminds them of the evil U.S. enemy that is trying to hold North Korea down. I could understand the immense pride the people of North Korea feel about their nuclear program, which in their eyes is a step toward becoming a self-sufficient, powerful nation.

"I do hope our two countries can become friends over time," I said.

"I believe you," Paris replied forgivingly.

⟶ **LISA**

IT HAD BEEN WEEKS since we'd heard anything about the activities Al Gore was engaged in. Every few days I'd shoot an e-mail to him and Kurt asking if they had heard anything. I kept getting variations on the same response: "no," "nothing yet," "no new news."

It appeared that whatever communication was under way with North Korea had gone dark. The discussions we thought were taking place seemed to have ceased entirely. Some of those advising me, with knowledge of North Korean etiquette, told me that nonresponsiveness was their way of saying no. In other words, it was becoming increasingly apparent that sending Al Gore to North Korea was not going to happen. This was perhaps one of the most frustrating periods of the whole ordeal.

If former Vice President Gore had been rejected, what could the North Koreans possibly want? We had reached an impasse. We were starting to feel doomed. My sister and Euna were the first Americans ever to be tried in North Korea's Supreme Court. They were the first Americans to be sentenced to serve time in a labor prison. Was it possible that they would be the first Americans to actually have to carry out their sentence?

Toward the end of June, an actress I know referred me to an international businessman of Chinese descent who travels in and out of Pyongyang regularly. I can't use his real name, because anything having to do with North Korea engenders suspicion. Here I'll call him Robert Hong. His résumé reads like a page out of a capitalist manifesto. He claims expertise in a plethora of areas, including tax consulting, gaming, telecom, media mining, financial services, property development, biotech, entertainment, and banking. He claims to hold official positions with four governments and is consulted by presidents, prime ministers, and other state leaders who value his advice.

I reached him by phone a couple of weeks before he was to make another trip to Pyongyang. He told me he would be flying there from Beijing for a series of meetings that would take place over the course of twenty-five hours. After several conversations with him, I still couldn't get a grasp of what he was actually going there to do. He told me he had some international investors interested in pouring millions of dollars into the Communist state. I asked him if he could help us. For a fee, he told me, he could persuade those holding the girls to release them—he was fairly certain of it. He said that doing it his way would yield far better results than if we continued to wait for governments to communicate. That, he said, would take forever.

"This is how you do things there," Robert said. "If you wait for the U.S. government, it will never get done. The situation is too bad right now. We should try to get them out quietly."

I didn't know what kind of money we were talking about, and I didn't ask. I figured that if he were successful, I would find a way to repay him, even if it meant that our family had to sell everything we owned.

On July 6, Robert sent me a text saying that he was leaving Beijing for Pyongyang and that he would reach me as soon as he got out.

~ LAURA

THE NEW MAN IN CHARGE, whom the judge referred to as my "guarantor," did not take walks with me outside, but I was allowed to walk along one side of the building under the supervision of my guards for thirty minutes each day. At first I misunderstood where the boundaries were and continued to walk along the perimeter of an off-limits area. My guards must have been confused as well, because they continued to let me wander. Suddenly I heard the guarantor shouting at me to turn around. Paris rushed over and told me I was not permitted in any area beyond the short length of one specific wall. My curiosity was piqued, and I became convinced that Euna was being held on the other side of the building. I pretended to cough loudly, hoping she might hear me and know I was nearby.

As I jogged back and forth within the permitted area, I noticed some workers installing a set of floodlights around the perimeter of the building. It seemed they were increasing the security of the area.

BEFORE THE TRIAL I had been receiving batches of letters every week or two. But under my new supervisors, I hadn't received any letters in almost a month. I was desperate for information from back home and news of any progress. I repeatedly asked the guards if there were any letters for me, but my inquiries went unanswered.

The discomfort in my abdomen continued to worsen. It felt as if knives were stabbing me in my lower stomach area. I threw up every bite of food that I ate, and a doctor was called upon to see me. The doctor was a gentle woman who appeared to be in her late fifties. She continued to conduct regular checkups on me once, sometimes twice, a week. I looked forward to her visits because they were an opportunity to interact with another person, even if it was just to talk about my health. I told her about my pains as well as my inability to sleep at night. She determined that I had a mild appendicitis and prescribed some antibiotics as well as diazepam, an older form of Valium, to help me sleep. The guards would administer the medication to me each day to make sure I wasn't taking more than the allowed dosage. Each night when the guards cranked up the volume on the television while watching the evening's Korean War flick, I popped a diazepam and drifted away.

Every morning, along with the antibiotics, I was given a packet of Emergen-C, an effervescent vitamin drink mix manufactured in California. I was surprised when I saw the familiar blue and yellow packaging. I had often taken Emergen-C back home whenever I felt the slightest onset of a cold. I wondered where the North Koreans had gotten this supply of the citrus-flavored powder, given the sanctions that barred any importation of goods from the United

States into North Korea. I felt bad that it might have come from an aid shipment and that it was being used on me, not on the North Korean citizenry who need it the most.

It had been a month since the trial, and for most of that time I'd been largely secluded. Then I was told that someone was coming to see me in the afternoon. I wondered if it was Mr. Yee. I hadn't heard from him since the last time we walked outside together, when he told me he'd try to visit me once a week. I waited anxiously, hoping he might be bringing some positive news from back home.

I heard footsteps approaching the room, followed by a faint knocking. The door swung open and I saw Mr. Baek standing in the entryway. I was happy to see him. Mr. Baek had always been kind to me and I missed his cheery disposition.

"Hi!" I said enthusiastically.

Though I could tell he was glad to see me, he was much more reserved than his usual self. After seeing whom he was with, I understood why he seemed so staid. Following Mr. Baek into the room was the prosecutor from the trial, along with the doctor and an older gentleman I had never seen before. The prosecutor wore the same imposing expression that never ceased to rattle me. He looked me over with the same indignant scowl that had greeted me on a number of occasions.

He began by saying it was his job to send me to the labor camp and that he was checking to see if my health had improved enough for me to go to prison. He questioned the doctor and asked for her assessment. According to the doctor, I was improving, but she thought I still needed a little more time for my appendicitis to heal.

"I thought that since you were in medical detention, your government might have done something to get you home before we sent you to prison," the prosecutor said. "But, it doesn't look as if they are doing anything. You should prepare yourself to go to the labor camp soon."

I was surprised to hear that no progress had been made. A month had passed since the trial, and I was hoping that what Ambassador Foyer had said was true, that the trial was a necessary part of the process, that the U.S. government needed a justification such as a long and unfair prison sentence in order to act.

"So, nothing has been done? There has been no word from my government?" I asked the prosecutor.

The older man sitting on the couch beside me began to chuckle. "Your government has been silent," he said. "Al Gore has offered to come here on a humanitarian mission, but he is the head of your company. Your government is trying to pass off your situation to your company rather than get involved. That is not acceptable."

His words felt like a sharp blade to the neck. I recalled the letter I had given to Ambassador Foyer weeks before the trial, which was meant for my bosses at Current TV. In it, I asked if Vice President Gore would agree to be sent to North Korea as an envoy. My suggestion was based on the conversation I had with Mr. Yee, who had acknowledged that Gore might be an acceptable representative. Now it appeared that he was not seen as a viable representative by the North Korean government because he was viewed as an extension of Current TV.

"Sir, I am the one who requested former Vice President Gore to come here. I believed he would be a great envoy, not because he is the chairman of Current TV, but because he is one of the most recognized political figures in the world. It's my fault. I did not know he would be unacceptable. If he is not the right person, just tell me who is, and I will try my best to make something happen. Let me call my family, and I will do whatever I can to get you what you want."

"Your family!" the prosecutor exclaimed. "All they are doing is complaining about your health and claiming that you must be released on humanitarian grounds. You had an ulcer before you came here. We did not give that to you!" He got up from his seat.

"Time is running out," he grumbled and headed out of the room followed by the others.

The prosecutor's words made me very anxious, and I feared that once the issues with my appendicitis were resolved, I'd be sent to prison. When the guard gave me that day's dosage of antibiotic medication, I pretended to swallow the pill but rushed into the bathroom and flushed it down the toilet. I hoped my appendix might actually burst, so that I would be taken to the hospital for surgery rather than being sent to a labor camp.

Later that evening, the older man from the prosecutor's office returned, accompanied by Mr. Baek. I was glad the prosecutor wasn't with them. The man asked me what I might say to my family if I was allowed to call them.

"Sir, tell me what needs to be done, and I will do my best to make it happen," I pleaded.

I was tired of trying to guess what the authorities were after. I knew I was being used to convey messages to the U.S. government via my sister, but no one was giving me any concrete information.

"I can't tell you what to do," he replied. "That would be a violation of your human rights!"

I refrained from laughing at this absurd statement. I could tell he wasn't trying to be humorous.

"Listen," he continued. "I work for the prosecution. It is my job to send you to prison. But I am also a father, and I sympathize with you. I am not speaking to you on behalf of the North Korean government. I am simply giving you some fatherly advice."

I didn't believe him. I knew that he, like Mr. Yee before him, had been sent to prep me for my next phone call with my sister. Lisa and I had become a channel through which the North Korean government was communicating with the United States. It was essential that I send the right message. But rather than telling me directly what they wanted in exchange for our release, the North Koreans preferred an indirect method. It seemed they didn't want it to appear that they had been feeding me instructions or making demands. In their eyes, we had committed a grave crime, and it was up to our government to make an apology and present a worthy envoy to mend the situation.

The older man said he believed an acceptable emissary must be someone who resonated with the Korean people: "The Korean people know about your crime. In order for them to forgive you, the envoy must be someone they recognize who can apologize on your behalf."

It was an interesting statement, given North Korea's totalitarian state and the mass propaganda machine that has brainwashed its citizens for the past six decades. Now they wanted a high-profile envoy to add to their propaganda.

"What about Governor Arnold Schwarzenegger?" I suggested.

I knew Kim Jong Il was a big movie buff, and I hoped he was a fan of *The Terminator*. I also recalled a letter Iain had sent that included a public statement from Governor Schwarzenegger expressing his concern after the results of the trial. As a California native, I thought it might be possible to get the governor of my home state to make the trip. The man discounted that suggestion immediately.

Still upset that Al Gore had been rejected as an option, I tried to convince the man that the former vice president was still the best candidate. "He's a Nobel Peace Prize holder," I said.

"What if you just cut off the 'vice' and go for 'president'?" he replied with a smirk. I was speechless. Was he really suggesting that President Obama was the only person who could secure our release?

"Sir, if you think President Obama is going to come here on our behalf, you might as well send me to prison right now," I responded, feeling defeated.

"I'm not suggesting the current president, but what about past presidents?" he replied.

I perked up instantly, and my mind began going through various options. I ruled out the two Bush presidents, thinking it would be more difficult to get approval from the current administration for their involvement. That left former Presidents Carter and Clinton.

"What about Carter or Clinton?" I suggested.

"Carter or Clinton," he said, mulling over the suggestions. "Those sound like good options."

I couldn't believe we were discussing men of such stature, but I was even more disheartened by how difficult the challenge might be to get either one to make the trip.

The older man got up to leave and told me he'd be back the next day. Mr. Baek followed him out, but returned shortly after and said he'd received permission to talk to me for a few minutes longer. He apologized for never getting a chance to say good-bye and asked how I was doing. I told him I'd been ill, and that I missed the old guards. I explained that I hadn't received any letters since his departure.

"No letters!" he said, sounding surprised. He told me he'd ask about the letters on my behalf.

"Oh, and happy belated anniversary!" he said, grinning.

I was touched that he had remembered. From all of Iain's letters mentioning our upcoming anniversary, I knew Mr. Baek was aware of our June 26 wedding date.

"Thank you!" I replied. "It's so kind that you remembered."

"Did they give you anything special for dinner that night? I told the woman in charge that it was your anniversary and to fix you a special meal."

I thought back on the meals I'd been given since the trial. I recalled one dinner that was different from what was normally served. It was a dark, pungent soup. I couldn't place what kind of meat was in the soup. The gamey flavor was too strong for my liking. I'd commented on its rich taste to my guard. "It's a special kind of soup," she'd replied. "Do you like it?"

Not wanting to seem disrespectful, I'd told her it was very good, while I concentrated on swallowing the tender bits of meat. I hoped it wasn't the Korean delicacy "sweet meat," better known as dog soup.

Later that evening, after Mr. Baek left, the guarantor brought me a batch of letters. They were the first I'd received since the end of the trial. I sifted through them, trying to find any sign of movement or news. I immediately went for the ones from Lisa and Iain and scanned through them for any information.

I gathered from reading the letters that very little progress was being made. In a note from Lisa, she explained how various non-governmental organizations, such as the International Red Cross, were concerned about our health. She also wrote that my doctor in the United States, Dr. Basil, had requested through North Korea's Permanent Mission to the UN that he be granted a visit to see Euna and me. Lisa also sought to convey messages to the North Korean authorities. She wrote:

> . . . Every day that goes by saddens me to no end. I was truly hoping that this could be a unique opportunity for our two countries to have some kind of meaningful exchange. I am still hoping that will be the case. I just hope it happens soon. It has been too long and your families miss you so much, sweetheart.

In a letter that was handwritten from Iain and then scanned, he wrote:

> Dearest Laura,
>     Things are moving very slowly at the moment. I don't know why that is. It is very frustrating, particularly for you

*and Euna. I wish there was something to do to speed things up. In the meantime we are pushing for a family and doctor visit.*

*I cannot believe another week has almost gone by. Another week without you. I am sorry, sweetheart. But don't worry we will keep on. I am thinking about you right now. I imagine your hair longer, being skinnier (from worry), but still strong and bright. Writing to you every day is the most important activity of my life at the moment.*

*Thinking of you every minute of the day.*

*Iain*

I could feel the anguish in their words and blamed myself for their pain. I was upset that weeks had gone by since the trial, and still things seemed to be in a state of limbo. But it wasn't the lack of progress that worried me the most. It was a news item Iain had included in his letter dated July 4 that filled me with unease. Sandwiched between a story from the *Economist* about the high court in Delhi ruling that consensual gay sex in India was not a crime and an excerpt about China delaying a mandate that all new computers be equipped with Internet-filtering software was one sentence about North Korea: "North Korea test-fired more short-range missiles, ratcheting up tensions in the region and defying recently tightened UN sanctions."

In the three months I'd been held captive, North Korea had conducted a satellite launch and a nuclear test, and was now firing off missiles. There was no telling what they might do next, including sending two American journalists to a labor camp.

I spent that evening obsessing over what I needed to say in my next call to Lisa. I thought back to the jokes Mr. Yee had made about me becoming more and more like a North Korean. In a sense, he was right. I wanted to get into the minds of the North Korean leadership so I could better understand how my government might best respond.

In addition to the mammoth request for an envoy, I hoped Lisa might be able to appeal to Secretary of State Clinton or President Obama to issue some sort of apology for our actions. I knew from the letters that various U.S. politicians had made statements after our sentencing. Some called for our release on humanitarian grounds and said the North Koreans should let us go without delay. But none had actually apologized for our actions. I could see that the North Korean authorities felt insulted by this perceived lack of respect.

I also planned on telling Lisa to cut back on the requests for medical visitations and to stop commenting on my poor health. The conversation I'd had with the prosecutor told me that the regime felt slighted by the accusations that my condition had worsened in captivity and that they hadn't been treating me well.

Throughout the night, I rehearsed in my head what I wanted to say on the call. I figured I would be given roughly ten to fifteen minutes, and I knew that each second was precious. The next day, the man from the prosecutor's office came to escort me back to the Yanggakdo Hotel.

Before making the call, the man sat me down and gave me these instructions: "You must tell your sister that this is a life-or-death situation. She needs to focus all her energy on getting you out of here. Your life is in her hands."

"Sir, with all due respect, I will not tell my sister that. I don't need to put any more pressure on her than what she is already under. I don't know if Carter or Clinton will agree to come here, but Lisa will spend the rest of her life working to get me home. That I know."

O N JULY 7, LAURA called again. Earlier that night I had spoken with Robert Hong, who had just come out of North Korea after his three-day trip.

"I don't have good news, Lisa," he said. "They said that your family has hostile intentions against their country, and they have made a decision that this situation has to be dealt with politically. There is nothing I can do about it."

Then in a grave voice he said, "I was told that the girls are ill, and their health is rapidly deteriorating."

"Oh my god, Robert!" I screamed. "What does that mean?"

"That your government better act soon," he solemnly replied.

When I hung up the phone, I was near hysterics. Paul came to comfort me, and I told him what Robert had just said about the girls' health.

"Babe," he said, "they're probably saying that to delay sending them to a labor camp. As long as they are ill, I don't think they'll send them away."

"They're punishing her for what I did," I exclaimed. "I have to get her out!"

I called my mom and Iain to brief them on Robert's report, and was just starting to calm down when the phone rang again. It had been sixteen days since Laura's last call, but when the phone rang at 10:15 P.M., I knew it was her.

The two previous calls had come entirely unexpectedly, but both had come shortly after 10:00 P.M. Since her last call, I had started making a point of being alert at that time every night.

"Baby Girl, are you okay?" I screamed breathlessly. I desperately wanted to know about Laura's health.

"Li, be calm," she said. "Just listen to me, okay?"

I scrambled for my notebook and a pen and put the phone on

speaker so that Paul could listen in as he had done during Laura's previous calls.

"Do not talk about my health in the press," she said decisively. "It will anger people. It will seem like our family is accusing the government here of mistreating us. I am being seen by a doctor. I'm okay."

I was confused by what Laura was saying because I'd just talked to Robert, and he'd said Laura's health was in trouble. So what was she saying now? Perhaps in anticipation of Robert's call to me, the North Koreans were trying to send a message through Laura that speaking publicly about her health would make the situation worse for her. The North Koreans seemed to be trying to ensure that no one believed they had abused the girls' human rights. Laura's tone was very deliberate, which seemed to indicate that her captors were talking through her. I knew from Laura's plea that the North Koreans must have seen some of the interviews our family had done, during which we expressed our concern for her ulcers and possibly deteriorating health. I was desperate to know the truth about her health, but she didn't give me a chance to ask about it.

"The feeling here is that our government doesn't care about us," Laura urged.

She said her interrogators were pointing out that America had not officially apologized and that some kind of acknowledgment of the girls' crimes had to be made publicly.

"Do you think that Secretary Clinton or President Obama would just say that they're sorry American citizens broke North Korean law?" she asked.

I thought about what such a request would mean diplomatically. President Obama had been in office for only seven months, and Secretary Clinton was under pressure from the six-party member countries—particularly Japan—to stand her ground against North Korea's provocations. Japan's leaders had loudly expressed their concern that North Korea's missiles could reach its country's population centers as well as the U.S. military bases there. I also knew that conservative hawks were monitoring the administration's every move and utterance, and were ready to jump on any signs of weakness or kowtowing to the repressive North Korean regime.

"I will try," I said, while keeping the geopolitical implications to myself.

"Okay. Li, we need to talk about an envoy," Laura continued.

"Vice President Gore has been ready to go," I said. "It was all set up and then everything went silent."

"Al isn't going to work," she replied. "They associate him too much with Current TV. The best thing he can do is work behind the scenes. But please thank him for everything he has done. I am so grateful."

"It has to be someone symbolic," Laura said. After a brief pause she continued, "Do you think either President Carter or Clinton might be willing to act as an envoy?"

I couldn't believe we were even having this conversation. We weren't talking about average Joe diplomats—these were former U.S. presidents! I knew that President Clinton was well regarded in North Korea from my trip there in 2007, but I immediately discounted him as a possible envoy because he was married to the current U.S. secretary of state. It could be perceived as an insult to Secretary Clinton if her husband played a role in this, especially considering her recent terse remarks about North Korea's nuclear ambitions. I didn't want to complicate things for her, because I knew she had taken our matter very seriously and I appreciated how much she wanted to bring the girls home.

I also wasn't sure how President Obama felt about former President Clinton, given the tension that had arisen during the presidential primaries when then Senator Clinton was running against then Senator Obama. And finally, I wondered how the Bill

Clinton option would go over with Vice President Gore. After all, Gore had spent months tirelessly trying to get the girls out. But the other option of President Carter left me a bit stupefied.

"Are you sure that Jimmy Carter would work?" I asked, surprised that his name had even been brought up.

I thought about the former president's age; I knew he was well over eighty and wondered if he would be able to undertake such an arduous and unpredictable mission. Regardless, I thought it was best to deflect attention from the Clinton option because it would be just too complicated. I nervously launched into a monologue about why Carter was the ideal choice.

"It's true President Carter is universally beloved," I declared. "He has been instrumental in the movement for peace and even won the Nobel Peace Prize for his efforts years ago."

"Yes, Li," Laura responded, "and he's been an important player in the Middle East peace process."

I could tell by her tone that she knew what I was trying to do and she was playing along.

We were nearing fifteen minutes on the phone when she asked me to jot down a list of things we should send by mail. Among the items she asked for were sunblock, feminine products, journals, and Clorox wipes. Sounding defeated, my sister told me that she was prepared to be sent to a labor camp. If it were to happen, she wanted to bring some items from home with her, as she would likely not have contact with the outside world for more than a decade. At the end of our conversation, Laura asked one more thing of me.

"Will you write a letter to Euna for me? Tell her I love her."

Laura's final request of me was confirmation that she and Euna had been kept apart. We still didn't know if they were being kept in the same location; all we knew was that both of them were inside North Korea.

After my talk with Laura, I called Michael to see if Euna had called him and if she'd said anything I needed to know. She had reached Michael, but other than conveying that the United States should apologize, she didn't say anything political or make any requests for an envoy.

## ᘰᘂ LAURA

Y HAND TREMBLED as I hung up the receiver. I replayed the conversation in my mind, wanting to be sure I had said everything I'd meant to say. It hurt me to hear the determination in Lisa's voice. She was so quick to promise President Carter. But what if he couldn't come, or if the U.S. government wouldn't approve of him? I didn't want Lisa to feel responsible if things didn't work out.

"Don't promise anything, Li. Just do your best," I had told her.

As I walked through the hotel lobby, my eyes bloodshot from crying, I saw a foreign tour group milling about. A young man who appeared to be from another Asian country was in the doorway, and he was staring in my direction. I wondered if he knew who I was from international news reports. I looked at him, hoping he might recognize me and perhaps contact my family after his trip to let them know he had seen me. I intentionally dropped my tissues on the ground as I passed through the doorway and bent down to pick them up. We locked eyes for a moment before my escort directed me to move along quickly. Nothing ever came from this brief encounter.

The next week was one of introspection. No one other than the doctor came to see me, and I was left with a lot of time to think and reflect. Up to this point I had largely kept any thoughts of going to prison out of my head. Now I decided it was time to prepare myself mentally. I thought about the estimated two hundred thousand political prisoners who are sent to the brutal Soviet-style gulags to be "reeducated" through hard labor such as mining, logging, or agricultural work. Family members of those accused of political crimes, such as saying something negative about the North Korean leadership, can also be sent to a prison camp. I told myself that if I were transferred to a camp, I would be enduring what many innocent North Koreans have had to struggle through for generations. Their stories of perseverance encouraged me to be strong. I considered myself lucky to have lived such a privileged life for as long as I had. Several times throughout the day, I sat cross-legged at the edge of my bed and meditated. With each inhale and exhale, I cleared my mind of any thoughts or fears, and for brief moments felt a sense of peace.

One afternoon I noticed Paris packing up some of her belongings. I knew the guards were allowed a break every six weeks or so when they could go home for an evening and spend time with their families. But they were not allowed to discuss the nature of their job with anyone. Paris told me her family believed she was working on a special assignment with a foreign tour group, not translating for one of North Korea's prized American prisoners. As she scrambled about the room, making sure she had her toiletries and cell phone, I asked if she was going to visit her family.

"Yes," she said hurriedly, "but I'll see you tomorrow."

"I'm really happy for you," I said. "I'm sure they miss you a lot."

She looked at me, smiled, and made her way toward the door.

"Have fun!" I exclaimed.

"Thank you. Just let the guards know if you need anything," she said and waved good-bye.

Without Paris there, I felt more alone than ever. The other guards treated me like an evil leper. Thankfully, I had a Sudoku puzzle book Iain had sent, and I used that to ingratiate myself with them. I ripped out a few sheets and offered them to one of the guards. She readily, but unemotionally, accepted them and immediately went about trying to solve the puzzles. Even if this small token didn't alter her attitude toward me, it did occupy her time so she wasn't consumed by shooting me harsh glares.

Paris didn't return for two days. When she did, she looked more energetic and refreshed. She had on a new outfit, a dainty pink skirt and a white blouse. She spoke exuberantly with the other guards in Korean. I assumed she was recounting her activities at home.

While I was eating my meal in the guards' room, I asked her about her visit. Suddenly her mood turned from glee to melancholy.

"I didn't want you to know I was seeing my family," she said, looking crestfallen. "But you asked, so I didn't want to lie. I felt bad that I was getting to see my sister and mother and father, when you have not been with your family for so long. That's why I told you I would only be gone for one day."

I wanted to leap from my seat and hug her like a sister. I was deeply moved by how considerate she was of my feelings. Unlike the other guards, Paris treated me like a human, like a friend.

### LISA

BRENDAN CREAMER HAD ARRANGED for vigils to take place all over the world on July 9, to mark one hundred days of Laura and Euna's detainment. Among some of the locations in which they were to take place were Chicago, New York, San Francisco, Orlando, Phoenix, Portland, and Seoul.

I had already planned on attending and speaking at the vigil in our hometown of Sacramento when Laura's call came unexpectedly two nights before. The original plan was to talk about our concern for the girls' health, but Laura had told me to end all discussion about her health, and that meant a change in strategy.

Of course, right after Laura's call I called Iain to compare notes. She had also asked him to see if the president or secretary of state would publicly apologize for the girls' transgression and if a former U.S. president would act as an envoy. I sent urgent e-mails to Al Gore and Kurt Tong after I hung up with Iain.

"Laura called," the e-mails read.

First thing the next morning, I briefed both men on my conversation with Laura. It was hard for me, but I broke the news that Gore was not seen by the North Koreans as the appropriate person to go to rescue the girls. I explained that Laura said he was not seen as suitable because he was the chairman of the company for which the girls worked. The North Koreans wanted someone more symbolic like President Carter or Clinton. Neither Al nor Kurt even acknowledged that Clinton was a possibility, and each was similarly surprised that Carter had been suggested.

Though a renowned international peace negotiator, Jimmy Carter had become known as a bit of loose cannon after veering from the official script during a number of diplomatic missions. The most glaring example had been a visit he made to North Korea in June 1994 when Bill Clinton was in office. Having been out of official office for thirteen years, Carter was asked by the Clinton administration to go to Pyongyang as a private citizen to talk to then leader of North Korea Kim Il Sung about its nuclear ambitions. Carter reportedly went way beyond his instructions by negotiating specifics of a nuclear nonproliferation treaty and announcing the terms live on CNN only minutes after he alerted Clinton to his plans.

Though the incident surprised members of the Clinton administration, it apparently softened the tone of the U.S.–North Korean relationship. As vice president, Al Gore had been front and center for the entire episode and recounted parts of it to me on the phone. He then got in touch with Carter to communicate

Laura's request. Even though Carter was eighty-four years old, he agreed to go. I reached out to people I knew with private airplanes. If a release were to happen, I didn't want it to be at the U.S. taxpayers' expense.

But I didn't want to get ahead of myself. We still needed to see if President Obama or Secretary Clinton would make some kind of apology. I pondered the implications. Could the six-party member countries see this as a concession? Would conservatives attack the president for legitimizing the repressive regime's legal system? I realized what a delicate dance such an apology would be.

My old friend Jeff Rose, a PR wiz in Los Angeles, strongly suggested that we publicly ask that our government request amnesty for Laura and Euna. We had been asking for a release on humanitarian grounds, but a call for amnesty would acknowledge the North Korean Supreme Court's adjudication of the case. Though we would not know the truth about what really happened until Laura and Euna were back, our goal was to get them home.

At the last minute, Iain decided to come with me to Sacramento to the vigil. It helped to have someone as studied and vigilant as he is to help me deliver our new message about amnesty. We had to convince the U.S. government that the onus should be put on the girls; they, not the U.S. government, had been accused and sentenced for crimes. Iain and I hit all the local Sacramento news affiliates, which were happy to have us on because that's where Laura and I grew up. Laura's story was big news there.

We took our message to the steps of California's capitol that Thursday evening, July 9. Beforehand, we briefed the legislators who spoke, including the person reading the statement from Governor Arnold Schwarzenegger, and asked them please to tread carefully in their remarks. These seasoned political hands might normally characterize the North Korean leadership as dictatorial and aggressive, but in this situation there was no room for error: the message had to be respectful and precise. At the end, hundreds of people chanted, "AM-NES-TY, AM-NES-TY," in unison. It was a moving sight that was picked up by national media outlets that night and the following morning.

By Friday afternoon the next day, news broke that Secretary Clinton had asked the North Korean government for amnesty on behalf of Laura and Euna. In an address to her State Department colleagues (that we later learned was set up for this purpose), she said that Laura and Euna had expressed "great remorse for this incident" and called on the North to allow them to return home to their families.

"We call on the government of North Korea to grant amnesty to the journalists," she said. "I think everyone is very sorry that it happened."

This was huge. To say we were shocked that Secretary Clinton acted so soon would be an understatement. I immediately called the president of CNN, Jon Klein. I had known Jon since my early twenties when I was with Channel One News and he was at CBS. I had recently worked on a big project for CNN called "Planet in Peril." Now I needed his help. If Laura and Euna's captors were watching CNN, I wanted to make sure they didn't miss this.

"Jon," I begged, "can you blast Secretary Clinton's request for amnesty all over CNN?"

He graciously obliged, and by the end of the day—morning time in Pyongyang—Secretary Clinton's apology and request of North Korea's leadership was being reported almost every twenty minutes on both CNN and Headline News.

I felt so grateful to have such contacts in the media. Working in the television business for more than twenty years had given me a wealth of relationships that were hugely helpful to our cause. I thought about the many people incarcerated overseas without political or media connections. I vowed to myself that if we were successful in getting Laura and Euna out, I would try to lend my voice when appropriate to support others who were being unjustly held by other country's governments.

Now we had successfully carried out part of what Laura had asked for; America's most senior diplomat had expressed regret for the actions of its citizens. The next step was getting approval for a visit by Jimmy Carter as an envoy to negotiate the girls' release.

# the envoy

∽∾ **LAURA**

**M**ORE THAN A WEEK went by after my call to Lisa and I'd heard no news. One afternoon Paris handed me another batch of letters. "I hope there is some positive information for you in here," she said sincerely.

I tore into the manila envelope from the Swedish Embassy, and to my astonishment, for the first time there seemed to have been actual progress and positive developments.

In a letter from July 18, Iain wrote:

> *Dearest Laura,*
>     *Waiting, waiting, waiting. I hope we don't have to wait too long. I wonder what is holding things up. As I said, your two main points are done. I am not sure how much more we can do . . . The longer it goes on and nothing happens the*

*more we need to look for other avenues. We have the apology.*
*We have the second step. I hope your assessment was correct.*
*I miss you too much.*

Though Iain wasn't specific, I gathered that either President Carter or Clinton had accepted the request to serve as an envoy. Not only that, he also said that Secretary of State Clinton had apologized for our actions and had asked the North Korean government to pardon us. I burst into tears and read the letter over and over, wanting to make sure the words were real. I told myself to temper my expectations, but I couldn't help feeling an overwhelming sense of excitement.

I rushed into the adjoining room where Paris was intently focused on her Czech language studies. "I think it might be happening!" I rejoiced.

Not fully hearing me, Paris took her headphones off her ears and looked over at me.

"I think either President Carter or Clinton may be coming!" I said. "I can't be sure, but my husband said that everything I asked for has been done. They're just waiting for your government to react."

"That's really great," Paris replied, smiling. "So now you are happy."

"Well, I don't want to speak too soon," I said. "All I know is that my government has acted. I just hope your government accepts their offer."

That evening I was visited by the man from the prosecutor's office and Mr. Baek. I felt certain they must be coming to relay some good news.

*"Ahn-yong-ha-sib-nee-ggah,"* I said, using the Korean greeting for "hello."

The man nodded in acknowledgment and took a seat. I tried to keep from grinning.

"Tell me," he began with a frown, "why did you insist on President Carter during your call with your sister? Who told you to ask for Carter?"

I didn't understand why he was being so stern. Surely he knew who had told me to request Jimmy Carter. He had. But before I could remind him of our lengthy conversation, he blurted out, "Carter, Carter, Carter! All you talked about was Carter! And you acted as if you were speaking on behalf of the DPRK government! You have upset many people by asking for Carter."

Stunned and speechless, I tried to gather my thoughts. "Sir, we talked before the call about Carter or Clinton. That's why I asked for him. I may have emphasized President Carter because I thought he would be more likely to come, and from the letters I received, it seems that he has offered to come. That's what we discussed here in this room. Why is there a problem now?"

He went on angrily about why Carter was not acceptable. He explained that Carter had been out of office for too long. Then he said it was up to me to figure out what I needed to do to go home, and that my next call to my family would probably be my last.

Unable to control my emotions, I began to bawl hysterically. For the first time, I raised my voice in frustration.

"You and everyone else listening in on my call knew I asked for President Carter," I said. "That was well over a week ago. My government has responded, and he has offered to come. Now you're telling me he's not the right person? Why couldn't you tell me earlier, right after the call? Then my family and my government wouldn't have gone through all the trouble! Do you know how difficult it is going to be to ask for someone else? And who is going to be acceptable? President Clinton? Or is it another person now?"

I didn't care if I was being disrespectful. I was tired of this charade. The highest levels of the U.S. government were following my lead, and now I was going to have to tell them it was all a mistake.

The man told me I would have one more chance to win my freedom by making another call to Lisa the next day. He asked what I planned to tell her. I wanted to make sure that this time I asked for only one envoy. Bill Clinton seemed to be the person the North Koreans wanted, but I needed to be certain. I went through a list of names from former Secretary of State Colin Powell to Christopher Hill, the U.S. ambassador to Iraq, who in 2005 was the head of the U.S. delegation to the six-party talks and had traveled to Pyongyang on a diplomatic mission in 2007. But everything came back to Clinton.

"I am just giving you my personal advice," the man began. "I think you should tell your sister that Clinton is your best and last option."

Those words solidified my decision. It had to be Clinton. As the man and Mr. Baek walked out of the room, Mr. Baek turned around and said to me in a hushed voice, "Try hard to get Bill, Laura."

Even he knew it was Bill or prison.

I thought about the likelihood that the former president would be willing to act as an envoy and the political hurdles that would have to be overcome for Washington to approve his visit. I knew Lisa had contacts who could reach Bill Clinton, and I was confident that Al Gore would be influential. But I also wondered if the secretary of state would want her husband to make the trip. There were too many factors to consider. My head was spinning. Instead, I decided to focus on my part and my part alone—the call. I spent the night scribbling down notes in preparation. I walked circles about the room, talking to myself out loud as if I were already talking to Lisa.

The next day I was taken back to the Yanggakdo Hotel to make the most important phone call of my life. This time, the only people accompanying me were Mr. Baek and Paris. During the car ride over to the hotel, I asked Mr. Baek, "Do you think the man from the prosecutor's office is speaking for the government? I know he says he's just giving me his personal opinion, but he must be sending me a message, right?"

"You know I can't answer that, Laura," Mr. Baek said. "But I do think he is very smart and knows what he's talking about."

"Thank you," I said. "I just want some reassurance that they won't ask for someone else later."

Pulling up to the hotel entrance was beginning to feel routine. The same hotel worker always opened my door and proceeded to escort us up to the second floor where there was a series of small conference rooms. No matter which room I was in, they all looked identical. Most of the space was filled by a rectangular wooden table with four chairs on either side. There was a calendar on one wall showing idyllic patriotic scenes such as a handsome soldier and a beautiful woman set against a sunset. Portraits of Kim Il Sung and Kim Jong Il adorned the main wall.

With tissues in hand, I took several deep breaths before picking up the phone and carefully dialing Lisa's number. After several rings, it went to her voice mail. My heart sank. I left her a message and proceeded to dial my mom, hoping I would be able to call Lisa again.

### ⌬ LISA

AFTER SECRETARY CLINTON'S REQUEST for amnesty, we had hoped for an immediate response, but ten days had gone by with nothing. Though we had also grown accustomed to waiting, every hour seemed to go on forever. We had another of our Friday conference calls with Kurt and the State Department team. Kurt confirmed that Carter's name had been floated to their sources in North Korea over the past few days, but there had been no response. It was exactly four months to the day since my sister was arrested. If what Laura communicated to me was true, Jimmy Carter could be on a plane within days. We seemed to be getting closer and closer, after four long months, and we felt more confident than ever that Laura might soon be coming home.

Several nights after the Carter option was presented to North Korea, Paul and I had just returned from a late dinner out. I threw my bag on the couch and rushed into the bathroom because I'd needed to go during the entire car ride home. From the loo, I heard my cell phone ring. It was inside my purse.

"Babe, your phone!" Paul yelled.

My heart sank. Anytime a call came after 10:00 P.M., I panicked, thinking it might be Laura. But I wasn't expecting to hear from her so soon after her last call less than two weeks ago.

"Answer it, damn it!" I screamed. "It might be her!"

The ringing stopped before Paul could get to it.

I knew it was Laura. I ran out of the bathroom, went straight for my bag, and dug out my phone. I was right.

"Li, it's me," said Laura's little voice on the voice mail. "I'm sorry I missed you. I hope they let me call you back, and I hope you are home. I love you so, so much."

I let out a scream and started wailing ferociously. I had been so religious about keeping my phone right next to me for months. Paul held me as I blubbered into his chest. It was only the fourth time my sister had been able to call me in four months, and I was so mad at myself for missing it. I dialed Iain, who didn't pick up the phone. He must have been on with Laura. I sent him a text message immediately: If you're on with her, please tell her to call me back.

**A**FTER NOT REACHING LISA, I carefully dialed my mom's number and was disappointed when it too went to voice mail. Then I tried my dad. No answer. Suddenly I was terrified that if I wasn't able to reach anyone, I might not be allowed to make another call. I was scared that my final chance to communicate with my family was disappearing with each ring. Finally I dialed Iain and heard his sweet, raspy voice on the other end of the line.

There was, for the first time, a tinge of optimism in his voice. I knew he and the rest of my family were just waiting for the North Koreans to accept Carter as an envoy. But it would just be a matter of seconds before I shattered what little hope he had.

I told him I had been wrong in saying that the North Koreans would accept President Carter as an envoy, and after thinking about it, I now realized it had to be President Clinton. Iain's response was filled with confusion, but I could tell that was a way of hiding his disappointment. Not wanting me to feel bad or to worry, he assured me that everything would be okay.

Iain's primary concern was that I might be transferred to another location or to the labor camp. As long as I was in my current location in Pyongyang, he could feel fairly certain that I wasn't being mistreated. He feared that once they moved us, it would not only be harder, if not impossible, for me to communicate with them, but it might be a sign that the window for diplomacy had ended. "You have to tell them that if they move you, it is going to change the dynamics, and it will be harder for us to get them what they want. We are working very hard on this. But if they move you, it's going to make things even harder."

"I think they've heard you," I said, referring to the people monitoring our every word.

"Yeah, I know, but just stress to them how important it is," he said desperately.

At the end of the call, I told Iain I was going to try to call Lisa again and that he should call her in about twenty minutes to compare notes. Hanging up the phone with Iain was devastating. I truly felt I might not get to communicate with him again for a long, long time.

Then I dialed Lisa's number, praying she would pick up. She answered immediately.

ᴖ LISA

I STARED AT THE PHONE for a half hour. Just after I asked Paul if he thought Laura would be able to call me again, the phone rang. It was her.

"Baby Girl, we achieved what you asked for," I excitedly explained. "And President Carter has agreed to go."

"Lisa, I made a mistake," she said, sounding defeated. "It has to be Bill Clinton."

I went ballistic. Did these people understand what it took to get these high-level people on board? Did they know what kinds of egos they were bruising—I was bruising? What kind of sick game were they trying to play? Frustrated by the enormity of the request and how often it seemed to change, I blurted out, "Who's telling you this?"

Paul elbowed me aggressively and covered the receiver. "You can't ask her that. You'll get her in trouble," he said gruffly.

After a slight pause, Laura responded, "I cannot speak for the North Korean people, but this is what I feel in my heart and in my gut."

I wanted to scream. I couldn't figure out if these requests

were based on Laura's whim or if someone else had dictated them. "Laura, your other option was just presented three days ago, on Wednesday," I explained, wondering if her captors might not have received the communication. "Did you know that?"

"If you haven't heard anything"—she paused—"it must mean something."

It was all starting to add up. This was how the North Koreans operated. They were masters of indirect communication. Early on, Governor Richardson's contacts seemed to entertain the notion of a visit by him, but then everything went dark. I had later learned that a detailed trip by Al Gore was actually presented, and then all talk of it ceased. I imagined that was about the time I got a call from Laura saying that Gore was too closely linked with Current TV. Both Richardson and Gore practically had their bags packed and were ready at a moment's notice to go and then silence. It became clear that in North Korea, they don't say no—they just say nothing.

Laura had made the request for Bill Clinton. But how could I know this was it? What if they kept upping the ante and asked for the secretary of state or even President Obama? There was no chance in hell that either would get on a plane to rescue two journalists inside a country that was taunting the United States and antagonizing much of the rest of the world. During my three previous conversations with Laura, I had made fervent promises that I would work my hardest to fulfill the requests she had made. I wanted my sister to know that she could count on me, and that I would be unwavering in my efforts to get her out. But I wondered if I was being taken advantage of by this bizarre regime. Every time "they" asked for something, we delivered. This time I had to act differently. I had to manage expectations. I would, of course, push relentlessly to see if Bill Clinton would go there, but I couldn't allow the North Koreans to think it was going to be easy.

"Laura, this is a very complicated request," I warned. "I just don't know if I can pull this one off, sweetheart. I don't know if you realize how hard this will be or if we can achieve it at all. He's the husband of the secretary of state, and what would this mean for Vice President Gore?" I added.

Then, in the most definitive voice she had ever used in all of our conversations, my sister replied, "Li, this is the best and last option."

Laura started to cry. I could feel the pain in her voice through the phone, and I wanted to reach through and pull her to me.

In closing, she uttered the following last words: "They say they are ready to send me to the labor camp, and I am mentally preparing to go. Please do what you can to see if President Clinton would be willing to act as an envoy. Make this your main goal. I love you so much."

I hung up the phone and gave myself a few minutes before calling my parents and Iain. Laura had not reached my mom or dad that night, only Iain, and she had conveyed the same message to him: it had to be Bill Clinton. I called Michael to see if Euna had asked for the same thing, and she had not. There had been no mention of Bill Clinton from her.

Iain and I discussed whether we should call Al Gore to let him know what Laura had just requested. Laura had already said that he would not be accepted because of his association with Current TV. I first sent an urgent text message to Kurt, who called me the next morning. I told him that Laura said she had messed up and that Bill Clinton was the last and final option.

"What about Carter? We just presented him a couple days ago," Kurt asked. I could sense the frustration in his voice.

"Laura said that if there is no response, that means something," I said.

He got it. He told me he would alert his colleagues in the

State Department but said the thing we all thought—"This one is complicated."

IKE IAIN, LISA WAS optimistic at the beginning of the call, and I hated taking that away with my news. I figured that she too had been guardedly preparing for our homecoming. But once I told her that President Carter was no longer an option, her confusion and disappointment came through in her panicked voice and hit me in the gut.

"First Al wasn't the right person, and now Carter. How do you know Clinton will be acceptable?" she asked desperately.

Borrowing the prosecutor's words, I said with conviction, "I strongly believe he's our best and last option."

On the ride back to the compound, I was glad I had a chance to speak with Mr. Baek, who had been listening in on my call. I asked him what he thought of the exchange.

"You did a good job," he responded encouragingly. "Are you sure you weren't really sent here to bring our two countries together?" He smirked.

Of all the people I had encountered in North Korea, Mr. Baek, who had met many foreigners over the years, was the most supportive of normalizing relations between our two countries.

I laughed. "Well, if everything works out, then I'll believe that things happen for a reason. But if I end up going to prison for twelve years, then I won't believe in anything anymore."

It was true. I did hope that our captivity might open the door to some forward movement between the United States and North Korea, two countries whose troops had been opposing each other for the past six decades and with a shared history of animosity, mistrust, and failed promises. But while I held out hope that President Clinton would come to Pyongyang to rescue us, my faith was beginning to diminish.

-⧉- **LISA**

**W**EEKS WENT BY AFTER Laura's call and plea for President Clinton. She had made herself very clear, and our family began pressing the State Department to get answers about whether the former president could go. It was such a challenging request that Kurt was no longer made aware of the conversations taking place about it. Laura's request was now in the hands of the highest levels of government, and it was being discussed among a small and exclusive circle of people.

In both our family conference calls and separate, private calls with me, Kurt said he had confirmed through his sources that the envoy had to be Bill Clinton. He told me that messages were being communicated from Pyongyang that they would wait as long as they needed to in order to get what they wanted. And the only thing they wanted was Clinton.

In my last call with my sister, she told me that the prosecutor was very aggressive and had been accompanying her to her medical evaluations. She said he was intent on making sure she was well enough to be sent to a labor camp. If she were indeed sent to the camp, hope of seeing her in the foreseeable future would be lost. No American had ever been sent to a North Korean labor camp before, but this was the first time one had ever been sentenced to serve in one. This was all new.

-⧉-

WE WERE NEARING THE END of July and then something came at me from left field. Through a contact who works with the Senate Foreign Relations Committee, I had heard some buzz that Senator John Kerry had volunteered himself to pay a visit to North Korea to try to bring home Laura and Euna. I reached out to the senator's senior East Asia policy adviser, Frank Jannuzi, who, to my surprise, confirmed that the North Korean leadership had accepted a private trip by the senator, independent of the State Department. They had even gone so far as to identify some possible dates in early August. Senator Kerry was trying hard to secure a commitment from the North Korean government that if he did travel to Pyongyang he would not return empty-handed. At the request of the National Security Council, Jannuzi had been coordinating exclusively with the White House on the possible Kerry mission, aware that separate, but parallel, efforts were also under way.

"Right now I am at a seventy-percent confidence level that our mission will be successful," he explained. "The White House wants iron-clad assurances that Laura and Euna will be returned safely. We cannot guarantee that yet, but I think it is doable and that we are making progress."

With permission from Senator Kerry, Jannuzi had first quietly reached out to Minister Kim Myong-Gil, the "New York channel," as early as mid May. At that time, Jannuzi told Minister Kim that Kerry would be willing to go to North Korea for two purposes: first, to secure the immediate humanitarian release of the girls; and second, to help create the necessary political environment for the resumption of six-party denuclearization talks. The talks had been stalled for months, and North Korea had conducted a long-range missile test on April 5, further chilling relations. According to Jannuzi, Minister Kim had thanked him for Kerry's interest but told him that "now was not a convenient time." As it turned out, North Korea tested a nuclear device on May 25.

I had never been alerted to this early action by Kerry's team or anyone else. As the chairman of the Senate Foreign Relations Committee, Kerry was uniquely positioned to represent an official imprimatur to a rescue mission, without actually being an official of the Obama administration. Jannuzi told me that Senator Kerry was "personally engaged in the issue" and had even called Minister Kim himself to impress upon the North Korean government the importance of allowing the girls to be released as a pathway forward in the U.S.–North Korean relationship.

During a follow-up conversation in June, Minister Kim told Jannuzi that Pyongyang was still considering a Kerry visit. Jannuzi retorted, "But don't invite us if you're about to do something else bad."

Then on July 4, America's Independence Day, North Korea launched a series of Scud-type ballistic missiles that provoked condemnation from surrounding nations as well as from the United States and the United Nations Security Council. A few weeks later, when things had calmed down after the missile test, Minister Kim personally called Jannuzi out of the blue to talk about a Kerry visit. North Korea's government had accepted Kerry's proposal and was prepared to welcome him to Pyongyang. And they wanted a visit to happen soon, as early as July 24, three days from then!

This was shocking to me, and I didn't understand what it all meant. First there was my sister's message—confirmed by State Department sources—that the envoy had to be Bill Clinton or no one. And now Jannuzi was telling me that not only had a Kerry visit been accepted, but the North Koreans wanted him to come immediately.

For a couple of reasons, Senator Kerry could not go to Pyongyang on that early date. First, there were important votes taking place in the Senate. The Senate recess did not commence until August 7 and he could not go in advance of that date. Second,

Senator Kerry was still trying to secure sufficient assurances from Pyongyang that his visit would be successful. Jannuzi asked me to keep all of this quiet, but how could I? If the North Koreans were ready to accept Kerry in three days and this was a window of opportunity to finally get my sister and Euna out of North Korea, I had to make damn well sure someone was on a plane.

I left a frantic message for Kurt on his cell phone as well as through e-mail. He was on assignment in Southeast Asia, and it was the middle of the night his time. He phoned me as soon as he woke up, which for me was the afternoon of July 21.

"Kurt, you know that I have held to a strict level of discretion throughout this process, right?" I asked. "Well, I am divulging some highly confidential information because of my sister's situation, but I have to know if this is true or not."

I told Kurt that someone in Senator Kerry's office had told me the North Koreans had accepted a visit from the senator and originally asked if he could be on a plane as soon as July 24.

"If this is an opening, if it's a window, we can't let it pass us by," I urged.

Kurt knew I was talking about Frank Jannuzi, but he didn't know that Kerry had been told he could come, and so soon. I told Kurt that Kerry couldn't go on the day the North Koreans put forward because of his obligations to the Senate. But, I wondered out loud, "If the North Koreans are saying now is the time for someone, maybe they want Clinton immediately."

"How can we expedite this, Kurt?" I asked. "Whom can I call?"

"I think you should call Gore," he replied. "He has more of a direct line to the White House than I do."

I hung up the phone with Kurt and called Al Gore right away. I again uttered the disclaimer about conveying sensitive information, but my sister was in danger and time was ticking away. Gore was surprised to hear the information about Senator Kerry. The Kerry route was a real tangent for those of us who had been following a certain path for so long. I could hear frustration in his voice. He had spent an exhaustive amount of time working on this and volunteering to go to Pyongyang. Now there were multiple names of acceptable envoys being tossed around.

I was again in the middle of a surreal situation.

President Clinton, Vice President Gore, Senator John Kerry, Governor Bill Richardson, these were some of the most powerful figures in American politics and society and here I was engaging them and their people regularly about taking a trip to the most secretive state on earth to try to rescue my sister. Plus, I was divulging private conversations concerning them. But I had to.

On July 22, I called Frank Jannuzi first thing in the morning to find out if he had learned anything new. He had. He told me that as of 11:00 P.M. the day before, senior officials at the National Security Council (NSC) had tentatively signed off on the Kerry visit.

"It's good news, Lisa," he exclaimed. "The White House is ninety percent convinced that we will return with your sister and Euna. They believe we are further along in our efforts to achieve success than their other options."

Jannuzi said they would be taking a doctor with them to check the health of the girls and that Minister Kim had confirmed an August 10 departure from the United States, with an August 11 arrival in Pyongyang.

But according to Jannuzi, the White House was still considering other options for an envoy. Of course, I knew of at least one of the people the Oval Office was contemplating, but I wasn't sure if Jannuzi or Senator Kerry had any idea that Bill Clinton was being considered. I had not revealed to Jannuzi that Laura asked specifically for Clinton and I never would because I had to keep that information top secret. Although senior NSC officials were backing a Kerry visit, Jannuzi said there might be other people who could get in sooner. Jannuzi didn't seem to know exactly who was being considered, but he said Kerry wasn't interested in obstructing a White House mission if someone else could go immediately and get the job done.

"We told the White House, if you can get someone on the ground sooner, do it!" Jannuzi said.

The John Kerry option was unexpected, but welcome nonetheless. Though the date being suggested, August 10, was later than we had hoped, it was an actual date. It was more definitive than anything else we'd had previously.

## ∾↩ LAURA

I LOOKED FORWARD to the thirty minutes when I was allowed outside each day. I would take in the fresh air, stretch my bones, and jog along the short length of the compound wall. I often closed my eyes and imagined I was back home, running along the streets of my neighborhood in Los Angeles.

On rainy days, my guards wouldn't let me go outdoors. I begged them to let me out so I could feel the cool drops on my skin, but they always refused. I'd never been a person who likes to get wet in the rain, but I yearned for it after being denied contact with the natural elements for so long. One day it began to drizzle while I was jogging, and the guards ordered me inside. I stopped right where I was, lifted my face to the gray sky, and let the droplets of rain mingle with my tears. Surprisingly, the guards stood back and let me have this moment.

The guarantor was really the only man I ever saw in the compound, but I knew others were around because at night I heard their booming voices singing karaoke from within the building. Sometimes, while I lay in bed, I could hear their inebriated voices yelling as loudly as they could. But it wasn't just the men; the female guards who were off duty joined in with the singing, and didn't come back to the room until well past midnight.

EVEN THOUGH I KNEW Lisa was trying to get Bill Clinton to serve as an envoy to rescue us, I was beginning to accept my sentence and resolve myself to being sent to prison. To prepare for this, I began working hard to keep up my strength so I could withstand whatever hard labor was required of me. I did extra push-ups, sit-ups, and yoga stretches each night. I meditated throughout the day, finding comfort in the sound of each breath.

But sleeping was always difficult. I eagerly accepted a sleeping tablet each night from the guard. When I tried to sleep without the aid of any drugs, I found myself restless and irritable. My mind would race with thoughts of my family. I worried about their health and mental condition. From the photos Iain had scanned onto his letters I could see that he was getting thinner. He had an emptiness in his eyes. I clutched onto his T-shirt at night as if it was a security blanket. Sometimes when I was able to drift off, I'd jolt myself awake moments later in a panic, searching to make sure his T-shirt was nearby.

-ಟಿಂ LISA

IT WAS HARD to imagine that things could become even more bizarre, but July 23 came along, and sure enough, they did. I awoke that morning to the news that an exchange had occurred overnight at a meeting of the Association of Southeast Asian Nations in Phuket, Thailand. Secretary Hillary Clinton was there, along with representatives from all the Asian nations, including North Korea. In reference to North Korea's defiance of the global community in its testing of a nuclear device, Secretary Clinton made this statement: "Maybe it's the mother in me or the experience that I've had with small children and unruly teenagers and people who are demanding attention—don't give it to them, they don't deserve it, they are acting out."

Even though many would agree with Secretary Clinton's characterization of North Korea's actions, her statement did not go unnoticed, nor was it left unaddressed, by the North Korean foreign minister, who was at the conference. He fired back by saying, "We cannot but regard Mrs. Clinton as a funny lady as she likes to utter such rhetoric, unaware of the elementary etiquette in the international community. Sometimes she looks like a primary schoolgirl and sometimes a pensioner going shopping."

Was I reading the logs of a primary-school principal or statements of diplomats at an international conference? A "pensioner going shopping"? What did that even mean?

The U.S. secretary of state had just been verbally belittled on the world stage by a tiny renegade country that was insisting behind the scenes that her husband was the only one they'd welcome to negotiate the release of my sister and Euna! Was this for real? What did it all mean? Did the North Koreans want to dig the knife in deeper by saying Bill Clinton or no one? Bill and Hillary Clinton may have been plagued by gossip and speculation, but one word was always used to describe their relationship: loyal. Would Bill Clinton even consider meeting with a government that likened his wife to a schoolgirl? It was both utterly absurd and devastatingly real. That's when I decided I needed to do whatever I had to do to get in touch with President Clinton. If I could just talk to him, I could convey how desperate we were to get my sister back.

~ LAURA

DESPITE MY EFFORTS to stay healthy, I developed some sort of stomach virus that prevented me from keeping down any food. It didn't feel like one of my typical ulcer pains. I was also stricken with an eye infection that caused the vision in my right eye to be blurred. When I came down with a high fever, the doctor was summoned along with a nurse. I felt like a pincushion as they began poking me with needles and injecting me with what they said was medicine to stop the nausea, reduce the fever, and calm my nerves. I was too weak to care. I could barely keep my eyes open. I was relieved to know they were trying to keep me alive and well. I took this as a sign that I was still worth something to them.

The vomiting had left me dehydrated and undernourished, and the doctor wanted to hook me up to an intravenous drip. The bottle of fluid needed to be elevated, so the guards went to work stacking a clothes rack onto a box, which was placed on top of a pile of books on a small table. The bottle was then secured to the rack with a sock. The rickety, makeshift contraption was an example of how, with limited resources, North Koreans have become adept at making do with what they have. I often saw the guards tinkering with the wires of broken extension cords, trying to get them to work for just a little bit longer. When the television set failed, they carefully adjusted different cables until the image appeared again, albeit fuzzier than before.

The nurse injected the fluid into my right arm, and the doctor wrapped me like a mummy in a thick blanket, hoping this would reduce my fever. In my delirium, I pulled my other hand out of the cocoon, feeling around for Iain's T-shirt. The doctor sensed what I wanted, reached over for the T-shirt, placed it in my hand, and put my arm back in the blanket.

I fell into a deep sleep for several hours. When I awoke, the doctor and nurse were in the same seats they'd been in when I'd drifted off. I smiled at them gratefully and thanked them for their help. Although I was dripping with sweat from being swathed in the thick blanket, my fever was still running high. The nurse replaced the IV fluid bottle with a new one and rubbed my arm to relieve the numbness.

"Do you get this kind of treatment in the United States?" the doctor asked me.

"We have very nice doctors like you," I responded.

"But if you get sick, do they come to your home, at any hour, any day of the week, and give you treatment?"

"No. Of course not. We have many problems with health care in the United States. It can be difficult to get treatment, and there are too many people who are unable to get the care that they need."

"In the DPRK, everyone receives treatment, no matter what," she said, beaming.

Paris chimed in proudly, "It's true. When I am sick, I can always see a doctor."

Like Mr. Yee, who often criticized capitalism for creating what he saw as a social and economic gap between the rich and the poor, Paris and the doctor were quick to denounce the U.S. government for not providing services to all. They seemed to find no fault with their own regime, which has left millions of people hungry. I wondered if they really believed all the propaganda about North Korea's perfect society.

A few hours later, Paris again commented on the good care I was receiving. "Wow!" she said. "You are being treated like the Dear Leader Kim Jong Il's wife. No one I know would get such care!"

As Paris and the doctor continued to extol the virtues of their health-care system, I thought back on Lisa's documentary about North Korea and the hundreds of people who showed up to be seen by a foreign doctor. Paris was right. I probably was receiving the kind of treatment that was unattainable by the average North Korean citizen.

The doctor and the nurse returned over the next two days to administer more fluids. I was so fatigued and incoherent that when Paris handed me another batch of letters from home, I could barely muster the strength to open the envelope. My eye infection had worsened, and I could barely see out of my right eye. Lying in bed, I sorted through the letters, found Lisa's and Iain's, and put the others aside to read later when I had more energy. I brought the letters close up to my face and concentrated hard on making out Lisa's words. "Do not be discouraged, there is a lot of movement in the effort to bring you home . . . Your request has proven to be quite complicated, as I thought it would be. Though we are still working very hard on it, it presents challenges on so many levels," wrote Lisa.

In a letter from Iain, he told me about a remark Secretary of State Hillary Clinton had made at the Association for Southeast Asian Nations Regional Forum in Thailand, where she likened the North Korean government to "unruly children." The North Koreans responded by calling her a "funny lady" who sometimes resembles a "primary schoolgirl." This was not good news for our case, but I didn't get angry or scared, probably because I was so drugged up that I was unusually calm.

I thought back on the four months I'd spent in captivity. There were nuclear tests, tightened sanctions, and now high-school-style name-calling. No wonder the two countries have been unable to find any common ground for decades. All I could do was laugh. I tossed the letters aside, as my laughter gave way to tears and I drifted back into a black daze.

RIGHT AWAY I REACHED OUT to a friend with close ties to Bill Clinton and asked if he could let the former president know about my sister's request. My friend agreed, and the next morning he called to tell me that Clinton was aware of the request.

"Tell Lisa that I will do my best to help in any way I can," my friend said, quoting Clinton.

I was told that the former president first wanted to talk to Secretary Clinton, who was returning from Asia in twenty-four hours. That same day, Al Gore phoned to tell me that he too had personally called Bill Clinton to ask for his help and that the former president had agreed.

"Let's just hope this is it," Gore said, clearly exasperated.

If the secretary of state was fine with sending her husband to North Korea, the one person in the world who could actually approve this visit was the current president. Through some of my contacts in the government, I learned that a few people in the Obama White House were concerned that a Bill Clinton visit might be too big a prize for a country that had so brazenly defied the United Nations. The United Nations Security Council's decision to punish North Korea for its renegade behavior was unprecedented. The secretive state had just been delivered a reprimand by the world community in the form of crushing sanctions. The White House seemed to be looking for ways to eliminate any possibility that adverse political consequences would result from such a high-level visit. I had also heard that President Clinton had not spoken with President Obama since the election, months ago.

The days that passed were agonizingly long ones. How could we convince President Obama to approve a visit by President Clinton to a country that was acting so unruly? A year earlier, I had met President Obama's sister, Maya Soetoro-Ng, and her husband, Konrad, when I spoke at La Pietra Hawaii School for Girls, where Maya is a history teacher. A mutual friend took the Ngs and me to dinner at a local Japanese noodle house. Maya and I hit it off right away. We spent much of the night talking about an issue we both feel passionate about: the sexual trafficking of young girls. I admired her intensity and her intelligence. During Laura's captivity, Maya and Konrad had checked in with me a few times to offer support. I never asked any favors of Maya, but I needed one now. I called to ask her if she could pass the following letter along to her brother, as he was the only person who could help us.

Even though a visit to try to negotiate the release of my sister and Euna had to be a private one, it nevertheless had to be approved by the commander in chief, especially because North Korea is a country with which the United States has no diplomatic relationship. We didn't for a second expect the U.S. government to pay for a trip to retrieve the girls. We just needed President Obama to say that it was okay.

Maya graciously agreed to send my note but suggested that I make it brief because her brother "has a lot going on," she said. *No kidding,* I thought, feeling guilty for adding more to the president's already crowded agenda.

*Dear Mr. President,*

*Please know that I would never ask Maya to pass along information to you unless it was imperative and urgent. To be frank: you are the only person on the planet who can help us. I know that a number of high level envoys have been presented to the North Korean government to secure the release of my sister and her colleague. As I am sure you're aware, they have all been rejected.*

*However, in the last week, since Secretary Clinton called for amnesty, one person's name has emerged—President Bill Clinton. First,* from my sister during a recent phone call that was monitored and likely contrived, and second, from State Department sources.

These sources have apparently reiterated twice their desire to have President Clinton and it seems apparent that they are ready to deal now. It is THE ONLY time the North Koreans have directly made a request. All other envoys had been presented to them. Furthermore, President Clinton told Vice President Gore that he would be more than willing to go.

You are probably aware that a visit by Senator Kerry has been accepted by Pyongyang. The North Korean Government offered a July 24th date to Senator Kerry but he can't leave before the Senate vote on your healthcare bill. Senator Kerry proposed an August 11 arrival date and was accepted. The problem is that the girls are not guaranteed to be released under his agenda.

Additionally, our fear is that the longer this goes on and given the deteriorating health of Kim Jong Il, the greater the risk that an all out power struggle could ensue and the country could descend into chaos. There appears to be an opportunity now to avoid these dire scenarios. Kerry's office told me that if the White House can get someone on the ground sooner and get assurances of a release, that "they should by all means do it immediately."

The last four months have been devastating, as I am sure you can appreciate. I know President Clinton is a complicated request, largely in part due to the person to whom he is married. But the signs that the NKs are ready to deal are more apparent than ever. Mr. President, our families beg you to approve sending President Clinton as an envoy to secure the release of my sister Laura and Euna Lee. This is a strictly American issue that desperately needs American/Barack Obama style diplomacy. It is a desperate situation but it seems there is an obvious way out. With terror in her voice, my sister said "If something is not done very soon—this week—we will be sent to a labor camp." Please help her avoid this, there's no one else who can.

My deepest thanks, Lisa Ling

More than a week went by after I sent the letter to Maya to send to President Obama. The ball, it seemed, was squarely in the court of the U.S. government. Every couple of days I would get an e-mail from Al Gore asking, "Any news?"

And I would respond, "No, not yet."

Conversations were being held at such a high level inside the White House that even Gore was not privy to them. Then, on July 30, Kurt called me to corral our families together for a conference call.

"It's good news," Kurt reported.

"We've gotten the go-ahead," he said in his official State Department voice. This must have meant that both Secretary Clinton and President Obama's national security team had given their blessing.

"The plan is for liftoff on August 4, sometime in the middle of the night," Kurt said. "If all goes well, the women will be back in the U.S. by August 5."

Every one of us on the conference call—Iain, Mom, Dad, Michael, and I—let out a collective and deafening "Yeah!!!" We had experienced so many highs and lows throughout the nearly four-and-a-half-month ordeal. This was the affirming tone we had been waiting for since the end of March. The possibility that I could have my sister back from captivity in less than a week was overwhelming; it truly was the best feeling of my life.

Kurt told us we had to stay quiet about the mission.

"We don't want to do anything that could jeopardize this," he urged. "The press must not know, or we could end up at square one again."

Kurt was right. If word got out, the blogosphere, as well as news and political pundits, would surely begin hypothesizing about a range of things. We didn't want any kind of speculation about a "deal" or anything to be said that wasn't true. Soon enough, President Clinton's private mission into ultrasecretive North Korea was sure to become one of the biggest stories around the world, and we couldn't do anything that would endanger it from going forward.

◌◌ **LAURA**

**M**Y FEVER BROKE AFTER the third day, and I was finally allowed out of bed. Ointment that the nurse had dropped into my eyes had restored my vision somewhat. I remembered the letters, and I had trouble imagining a visit from President Clinton given the recent exchange between Secretary Clinton and the North Koreans. I decided it was time to find out what rights I had, if any, as a foreign prisoner in North Korea. Up to this point, the North Koreans had seemed determined to show the world that they had a just legal system, and I wanted to put their image of themselves to a test. First I would demand that my family be allowed visitation rights, something I knew they had been trying to obtain through North Korea's Permanent Mission to the UN ever since my detainment.

I asked Paris to tell the authorities I wanted to see Mr. Yee immediately. Two days later, I was surprised when he and Mr. Baek walked into the room.

"I heard you've been ill," Mr. Yee said. "Are you feeling better now?"

I told him I was getting better, and I was very grateful for the doctor's care.

"I was told you wanted to see me," he said.

"Yes, well, I thought you would be coming to visit me once a week, but then I never saw you again. I'm a little mad at you," I said in a friendly tone.

He blushed. "I have a job. My bosses would be very upset if I spent all my time visiting Miss Ling. Anyway, I'm here now. What is it that you want?"

Before I asked him about my rights and whether my family could visit, I began with my standard question: "Have you heard any news?"

"It seems that something may be happening," he replied casually.

"'Something' meaning something good or bad?" I asked anxiously.

"You will find out soon enough," he said and got up to leave.

"But you just got here!" I cried. "I have other things I want to ask you."

"I will be back tomorrow," he said. "In fact, I think I will see you every day from now on. Get some rest."


⟡ **LISA**

**T**HE NEXT FEW DAYS were challenging because friends were calling to check in on us as they had been doing for months. We had no choice but to tell them we were still feeling hopeful but hadn't heard any news.

Another conference call was set up for Saturday so our families could speak with Bill Clinton before the Tuesday, August 4, departure. This time Al Gore joined in.

Both my parents, Iain, and I were at my mom's house on different phones when the call linking us to Clinton and Gore came through.

"I am honored to have been asked by you all and Al to help get Laura and Euna back," Bill Clinton said in his characteristic voice.

"Thank you, Mr. President," Gore remarked. "We appreciate that you have agreed to take the time to do this."

And in a moment of levity, Clinton jokingly retorted, "You know, Al, this means you have to go to Port-au-Prince with me."

President Clinton was referring to his recent appointment as special envoy to Haiti for the United Nations. Haiti is the poorest country in the Western Hemisphere, and its capital city, Port-au-Prince, has been plagued by instability and violence in recent years. This conversation took place more than five months before the devastating earthquake that paralyzed that country.

"Yes, Mr. President . . . ," the former vice president responded in over-the-top effusiveness. As if to say, *I know you're doing me a big favor.*

Hearing the former president and vice president on the phone together was exhilarating. They spoke in a formal tone—because others were listening in—but were respectfully familiar at the same time.

I took a deep breath and began to address President Clinton to thank him on behalf of our families. Then all of a sudden, in mid-sentence, my call dropped. "Oh my god!" I exclaimed, verbally assailing my cell-phone carrier.

Fortunately, I was immediately called back to rejoin the discussion. Iain had quickly jumped in where I left off, and expressed our profound gratitude to the former president and his wife, the secretary of state. Clinton then reiterated how important it was to keep his participation utterly quiet.

"I have every confidence that we'll get them home," he said, "but I cannot stress this enough: no one can know about this private mission."

After four and a half of the longest months of our lives, in a matter of days the forty-second president of the United States would be on his way to the most secretive nation on earth with the hope that my sister and Euna would be on his plane when he returned.

The days and hours that preceded takeoff were filled with lo-gistics coordination. When a former president is involved, there's no such thing as simply landing at Los Angeles International Air-port and going through immigration and customs. A good friend of Bill Clinton's, Andrew Liveris of Dow Chemical, volunteered its plane to transport Clinton and his staff from New York to Burbank and then back to New York. Another close friend of the former president's, Steve Bing from Los Angeles, donated the use of his private plane and staff for the trip to Pyongyang and back, as well as his hangar in Burbank, California.

The plan was for the plane carrying former President Clinton and his team to leave Burbank at approximately 2:00 A.M. Pacific standard time on August 3, with stops in Alaska and Japan for refueling.

Linda McFadyen phoned and told us to pack a little bag for Laura that would be waiting for her on the plane. I rushed out to the Gap to buy my sister a brand-new, clean outfit that she could change into: a soft brown V-neck sweater and a comfortable pair of cargo pants. She had probably been wearing the same clothes every day for months, and I wanted her to have something nice. My mom filled the bag with toiletries and Laura's favorite snacks, and we all included notes for her to read on the plane. I made sure she had makeup and tweezers in the bag so she could be prepared in case there were cameras when they arrived back home. I knew my sister would want to look presentable if that happened. I fig-ured she probably wouldn't be looking her best after spending nearly five months in captivity.

I returned to my mom's house to show my family what I had bought for Laura. I pulled the sweater and cute socks out of the bag, and turned to Iain to get his approval. He was almost giddy, like a little boy. Typically he's pretty self-controlled and unemo-tional, but I had seen a different side of Iain throughout this har-rowing ordeal. His love for my sister was unlike anything I had ever witnessed. I had never seen him cry before this, and the pain he exhibited from not being with her was excruciating. It was awesome to see him exuberant, and it warmed my heart deeply. After we finished putting Laura's bag together, Iain pulled me aside to ask me a question.

"Do you think she'll be the same?"

"I don't know," I responded. "All we can do is love her and give her time to decompress."

"Make sure to tell your mother that," Iain quipped.

I laughed. "Most definitely."

Two days before takeoff, Kurt called to say that the White House wanted just the family members to be present at the hangar when the plane arrived in Burbank; no friends, just family. I conferred with Iain and my parents, and we felt strongly that Vice President Gore should be there. Without alerting Kurt or the White House, I phoned Gore and asked if he would like to be with us on the ground to greet the girls, and if so, if he would say a few words. He had just as much of a reason to be there as we did. He had spent the entire summer working on getting the girls freed. I didn't want him to be left out of the reunion.

I called Kurt and told him I had asked Al Gore to be present when the plane arrived and to make some brief remarks. There was a pause on the other end of the phone line.

Before he could say anything, I said, "It just wouldn't be right if Vice President Gore weren't there, Kurt." And then I added, "If the White House has a problem with it, they will have to call him themselves. I'm sorry."

They never called.

We were told that President Clinton wanted to be on the ground in North Korea for as little time as possible. But the North Koreans were insistent that he stay for a minimum of twenty hours. Unfortunately for our side, all the leverage was theirs. Clinton's counselor and closest adviser is Doug Band. He became my main point of contact for everything that was happening on the plane.

As soon as they were "wheels up" from Burbank, Doug fired off an e-mail to let me know the plane was in the air. We were so excited that our friend Morgan took a photo of my mom, Iain, and me holding up an old GQ magazine with President Clinton on the cover. I e-mailed that to Doug to show the president. Steve Bing's private 737 was fully loaded with the highest technical equipment, so Doug and I were able to communicate the entire time the plane was in the air. Every few hours I got an e-mail from Doug: " . . . halfway to Elmendorf [Air Force Base, in Alaska] . . . 15 minutes to Japan . . . can't sleep, too anxious . . ."

Every time the red light on my BlackBerry flashed, we all rushed to check the latest update. Doug let me know when they landed in Tokyo to refuel, and as soon as the plane lifted off again, the news broke. And it broke big. A private plane with a former U.S. president on board doesn't touch down in another country without anyone noticing. The South Korean press somehow caught wind of the plane and its well-known cargo and filed an immediate report. The story was out and the headline read: "President Bill Clinton Is on a Secret Mission to North Korea to Negotiate the Release of the Two American Journalists."

# the rescue

~∽ LAURA

THE NEXT MORNING MR. YEE returned and began questioning me as he had during the investigation, asking me if I remembered the crimes I had committed. He wanted me to tell him again what my motivations were for working on the documentary. I told him what he wanted to hear—that I was trying to bring down the North Korean government.

"So, you understand, then, why you were given a twelve-year sentence, according to our law?"

"Yes," I replied. I wondered why he was bringing all of this up again, but then it became clear what he was trying to achieve.

"If you are allowed to go home, what will you tell people about your crime?" he asked. "There are people who will ask you if you were tortured or if we made you lie. What will you tell them?"

My heart was racing. Was he saying I might be going home?

Three days ago, I was hooked up to an IV drip, feverish and incoherent. My spirits were at the lowest they'd ever been. Now, could this misery finally be ending?

"I will tell people honestly what happened," I said. "People have shown me kindness here, and I will never forget that."

He then proceeded to tell me what I'd desperately wanted to hear ever since my apprehension.

"In about an hour, an important envoy from the United States will be landing in Pyongyang. If things go well with his visit, you may get to go home with him. If things do not go well, then you will not go with him."

He also told me that Secretary of State Clinton's apology and call for amnesty had been received positively by the North Korean government. Thankfully, he didn't mention anything about "unruly children" or "schoolgirls."

"This envoy that is arriving soon will most likely want to see you and Euna," he said. "So you should prepare yourself to see him shortly."

"That's unbelievable!" I shouted, grinning from cheek to cheek. "Who is it? Is it President Clinton?"

"I cannot tell you who it is," he replied.

"But please," I begged. "What does it matter if I know now. He's already on his way."

"All I can say is that it's one of the people you asked for."

He then told me what he thought I should say to this envoy. "You should first tell him that you are very sorry for your crimes and that it's important that he apologize on behalf of the U.S. government for your actions and promise it will never happen again. You should then tell him that you really want to go back with him when he returns."

"That will be easy," I said.

I couldn't believe that everything was happening so fast. I was shocked. I wondered if that was why the doctor had cared for me so intently, so I would be healthy enough for the visit from this envoy, whoever he was.

Mr. Yee then got up and told me we were going for a walk outside. I leaped up and followed him. It had been nearly a week since I'd been out of the dimly lit room, and I winced as the bright rays of the sun hit my face.

As we walked, Mr. Yee brought up Euna. "You haven't seen Euna since the trial," he said.

"When will I get to see her?" I asked

"You will see her soon," he said with a half grin.

He proceeded through a door on the other side of the compound, a part of the building where I'd never been allowed to go. We walked down a short hallway into a corner room—one that was exactly opposite my side of the building.

Standing before me was Euna. I raced toward her hysterically, and we embraced. It hurt me to feel her thin, frail frame. She had lost a lot of weight over the months. Her cheeks were sunken in and pale. We cried in each other's arms. I didn't want to let her go, for fear we might be separated again.

There was another man in the room—he was Euna's interrogator. He was familiar to me because I had seen this tall, thin, professorial-looking man on a few occasions through the window in the guards' area. Sometimes, if I was too close to the window when a person passed by outside, the guards would scold me and tell me to move away from the area, but I had seen this man.

As I suspected, Euna had been in the same building all this time. Mr. Yee told us that we could remain together, and after lunch we were going to meet the special U.S. envoy.

Seeing Euna was like a dream. It was hard to believe that during the agonizing months of our confinement, we had only been together for a total of six days. Every day for the past four months, I would wake up and pray for her well-being. "Lord, please give Euna the strength, courage, and wisdom to get through another day," I would say out loud. My prayers had been answered.

Euna and I sat in her room and began to compare our experiences in captivity. We talked about the interrogation process, each worried that the other had divulged certain bits of information.

"I pretended not to remember Pastor Chun's name," I said, "but after a few days of questioning, they said they already knew who he was because you had confessed."

Euna denied doing any such thing. "I thought it was you who revealed his name," she said.

Both of us had figured the interrogators were pitting us against each other, but we were each determined to protect our sources. I hoped that nothing bad had come to the people who had helped us or opened their lives to us.

We shared what we had learned from our letters about the outpouring of support from around the United States and all over the world and how the thoughts and prayers of so many people gave us the strength to endure.

For lunch, the guards brought out an elaborate meal of cold noodles, fresh fruit, and pastries. My stomach was still weak, so they served me a bowl of gruel, which had become my staple over the last few days. But having not had much fresh fruit in months, I slowly nibbled on some pieces of melon, savoring the sweet flavor.

Paris came over to give me my medication, which I'd been taking with each meal. "This is my friend Euna," I said to Paris, who smiled.

"So now that you're finally with your friend, you've already forgotten about me," she said jokingly.

"Of course not!" I replied. I was touched that Paris felt a certain closeness to me.

"I'm just joking," she said. "I'm glad you can be together." She said we would be leaving soon to meet with the U.S. envoy.

My stomach was churning from being ill and also from nervousness. I continued to contemplate who it could be that was coming to our aid. I knew that Bill Clinton was the person the North Koreans wanted most, but maybe they had decided to accept Jimmy Carter after Secretary Clinton's harsh remarks. On top of that, none of the recent letters I'd received indicated that any progress was being made on the Clinton front, but those letters were at least a week old.

"I have a feeling it's either President Carter or Clinton," I said to Euna, "but I'm not sure which one."

After lunch, Euna and I were taken in separate cars to the Koryo Hotel, a twin-towered building built in the 1980s. When we arrived, our interrogators escorted us up an escalator to the second floor. They left us in a small conference room with the guarantor and Paris. I was so jittery and my stomach was so full of unease that I had to request to use the toilet every few minutes. The guarantor seemed worried that my condition might prevent me from being able to meet with the envoy. He offered me some stomach medication, but I declined, fearing it might make my stomach even worse. I closed my eyes and began to meditate. *Just be calm,* I thought. *It's all going to work out. You'll be home soon.*

"Euna, I think it's happening. I think we're going to be going home," I said. "But let's not jinx anything."

We held hands, waiting to meet our savior.


◦⟨⟩◦ **LISA**


**W**HEN THE NEWS ABOUT the Clinton mission hit the airwaves, our phones started ringing like crazy. Press from all over the world was calling to get a comment about what was happening. Still under strict orders to not speak, none of us answered calls from numbers we didn't recognize.

With TVs blaring, we were practically dancing around my

mom's house. I looked at Iain, who had lost a significant amount of weight during the ordeal, and smiled at the thought of him and Laura finally having meals together in their new home. In one day, years had been lifted from my mom's face. Dad was helping her prepare Laura's favorite soup: Chinese watercress. They were like an old married couple snapping at each other to pass the salt, but it was a joyful bickering—their little girl was coming home.

A blocked number kept appearing on my cell phone over and over again. Then an e-mail appeared on my BlackBerry that read "POTUS [President of the United States] is trying to call you from the Situation Room. Pick up the phone."

It was August 4, President Obama's birthday. My mom, dad, Iain, and I picked up four different phones in the house so we could all be on the line.

"Michelle and I are so happy that this day has come," President Obama said in his iconic voice.

"Thank you so much, Mr. President," I said. "We know that this was a very complicated situation for all involved, and we're so grateful for your blessing."

"Listen, I've been on this for a while," he answered, "and this was before I got the e-mail from my sister."

We all graciously thanked the president for taking our matter seriously despite everything that he had going on in the first months of his presidency. At the very end of the call my mom blurted out, "Happy Birthday, Mr. President!"

"Thank you," President Obama replied. "This has been a great gift."

Bill Clinton's chief, Doug Band, and I were e-mailing news reports back and forth. I told him that President Obama had called us to tell us how happy he was that this mission was happening. The plane was so wired that everyone on board was getting real-time news reports of the Clinton trip, so Doug knew everything. And then all of a sudden, communication stopped. I knew at once that meant President Clinton and his team had landed in Pyongyang. I recalled my trip to North Korea, when my cell phone was seized immediately upon arrival. I imagined that while on the ground in North Korea, the Americans might not be able to communicate freely until they were back in the air.

Approximately two hours after I lost contact with Doug, photos started to emerge from broadcasts on North Korean television of President Clinton's arrival inside the Communist country. He was shown coming down the stairs and onto the tarmac, but as he reached out to shake the hands of the North Korean officials, he was wearing the most expressionless face the world had ever seen on him. I would later learn that persons in the White House and State Department suggested that he not appear too affable under the circumstances, and that he had practiced maintaining that look of total stoicism. On the ground to greet President Clinton was North Korea's chief nuclear negotiator Kim Kye-Gwan. For months we had been urging all parties to keep our issue separate from the nuclear one, but on this day it seemed as if the North Korean leadership was trying to make some kind of statement by having Kim Kye-Gwan there to meet with the former United States president. We wondered if this meant there might be an opening for discussions about nuclear disarmament in the future.

We were glued to the television; it was the biggest story of the day. We had been able to keep the mission secret for days, but now it was everywhere. We wondered if President Clinton had seen Laura and Euna yet—or if they even knew he was there. What was going on inside North Korea?

**A**BOUT AN HOUR LATER, we were told that the envoy had arrived. We were ushered out of the room and led down a long corridor. The path was lined with at least twenty North Korean security agents dressed in black. Their expressions were stone cold and intimidating. As I made my way down the hall, all of a sudden, at the end of the line of North Korean officers, I spotted a single bald-headed American, wearing an earpiece. It was a U.S. Secret Service agent. Seeing him there gave me goose bumps. I could feel the presence of my country standing before me.

When we reached the end of the corridor, two doors swung open, and standing ten feet in front of us was President Bill Clinton. Perhaps it was the way the room was lit, or my overwhelmed state of mind, but I felt like the former president was shrouded in a bright, beaming light. In my eyes, he looked like an angel who had come to our aid. I was awestruck. Every single moment of my captivity had felt utterly surreal, and this was no different. President Clinton had traveled halfway around the world to Pyongyang to rescue us. It should have been a scene out of a movie, not out of my life.

Unable to control our emotions, Euna and I burst into tears as we stepped toward the former U.S. president. He looked at us with fatherly concern and embraced us tightly.

"Bless you, bless you," he said in his smooth southern drawl. He spoke in a hushed tone, out of earshot of the half a dozen or so people in the room. North Korean photographers and videographers seemed to be recording his every move.

"President Clinton, thank you so much for coming. You were the only person who could save us. We are so grateful," I said. I then reiterated what Mr. Yee had instructed me to say.

"Well, that part has been done," the president responded, referring to making an apology on our behalf.

He told us that he had just come from a good meeting, though he didn't specify with whom. He also said there was still a little more work that needed to be done, but he felt confident that we would be leaving on a plane with him and his team the next morning.

"I'd like you to speak with my physician about your health and whether you can fly," he said and motioned us to his doctor, Roger Band.

I knew from Iain's letters that the swine flu virus was spreading in the United States and other countries. Iain joked in one of his letters that the one thing he didn't have to worry about was my catching swine flu, because North Korea was so isolated. Even so, I didn't want to take a chance by telling Dr. Band that I had just gotten over a serious fever. Nothing was going to keep me from getting on that plane.

"We've been treated fairly," I told Dr. Band. "We might have a few ailments here and there, but it's nothing that being on U.S. soil can't cure."

As we were talking, Doug Band, Clinton's top adviser, John Podesta, his former chief of staff, and Justin Cooper, his top aide, introduced themselves. Also part of President Clinton's team were Stanford professor and expert on North Korea David Straub, Min Ji Kwon, an interpreter from the U.S. Embassy in Seoul, and a contingent of seven U.S. Secret Service agents. The former president then came over to us again and said that he and his team were going to have to leave.

"Bless you," he said again compassionately. "We'll see you tomorrow."

After the room emptied out, our interrogators rushed toward us and asked us what Clinton had said. I explained that he still had some more work to do, but he hoped we'd be going home with him.

"Is that it?" Mr. Yee asked.

"Yes, that's it," I replied. "I hope it's true."

We went back to the compound, and I was told that I would be able to spend the night in Euna's room. There seemed to be a lot of buzz going on. Though we were confined to Euna's quarters, I could hear people moving about all over the building. There was a kind of frenzied energy in the air.

Euna and I contemplated the likelihood of our being released the next morning. Though we felt close to certain that things would work out, there was still a chance that something could go wrong. We were in North Korea, after all, a country notorious for being duplicitous. We tried to temper our expectations, not wanting to get our hopes up only to find out that we would not be going home. I tried to imagine what sorts of conversations the former president and his team were having with the highest levels of North Korea's government. Would President Clinton be meeting with Kim Jong Il? By nightfall, we still hadn't heard anything from our interrogators.

Since I was going to sleep in Euna's room, I asked the guard if I could go back to my room to collect my toothbrush and other belongings. We had to walk outside in order to get to the other side of the building, and I noted that this was the first time I'd been allowed outdoors at night. I looked up at the moon and the glimmering stars, sights I hadn't seen in months. I thought of my family and how anxious they must be feeling, knowing that our big chance to be released had finally come. The next time I look at the moon, I thought, I might be seeing it from home.

·❧· **LISA**

SOON MORE PHOTOS STARTED to surface. This time they included images of a man thought by the world to be on his deathbed, the infamous leader of North Korea: Kim Jong Il.

The Dear Leader was smiling and jubilant as he greeted America's forty-second president, who held on to his inscrutable expression throughout the trip. From an intelligence perspective, it was tremendously valuable for an American to see the notoriously despotic ruler. Every one of my advisers had been convinced that Kim was not well. He had not been seen publicly in a very long time, and whatever photos existed were said to have been taken years ago and doctored to make them look more current. Some bloggers even speculated that the reason Clinton had been requested was that the North Korean regime was going to perform a succession ceremony that would usher in a new era of leadership, and Kim Jong Il would hand over power to his son.

But judging from the pictures, there was no denying that Kim Jong Il was alive and well. President Clinton later told me that North Korea's leader not only was alert but was firmly calling the shots inside his country. He added that the younger Kim was not even present in any of the meetings during this visit.


∽∾ **LAURA**


T HAT NIGHT EUNA AND I lay in bed still unsure of our fate. The guard in the adjoining room was watching the evening newscast. Suddenly we heard a female North Korean newscaster say, "Clin-ton!" with the booming formal cadence I'd become so familiar with. Euna and I popped out of bed and rushed into the guards' room.

The news anchor was describing a meeting between Kim Jong Il and President Clinton. Then photos from the visit were displayed on the screen. The first photo was a group shot with Kim Jong Il and Clinton's team, all of whom, including Kim, had on serious expressions. It was hard to gauge the mood in the room. But the instant I saw the picture of Kim Jong Il with his wide, toothy grin, standing proudly next to the solemn-looking Clinton, I knew we were going home.

"The reporter said it was a warm meeting," Euna explained.

"We're going home," I said, no longer worried about jinxing anything. "We're going home!" We spent the rest of the night sharing in our excitement.

At 4:00 A.M., a guard came into the room and told us to get ready, that our interrogators were coming to see us. The guarantor brought in several boxes of books and things sent from our families that we had not been allowed to have. He gave us some duffel bags and told us to pack what we wanted. The boxes contained protein bars, dried fruit, shampoos, lotions, beef jerky, tissues, toothpaste, deodorant, and other basic items. I recalled how much I had craved and begged for the protein bars and toothpaste but was not allowed to have them because the guards were worried I might be poisoned. I left the items in their boxes. I thought they would be put to better use by the people there.

I threw a few articles of clothing and other small items into the bags, including the black lined boots I was wearing on the day of our apprehension along the border. I remembered that morning and wondered if circumstances might have been different had those boots not been so heavy.

The things I wanted to take home most, my prized possessions, were the letters I had received. I don't think I would have gotten through that terrifying time without the knowledge and wisdom contained in them. There were thirty manila envelopes in all from the Swedish Embassy, each containing several letters. I carefully packed the envelopes and all the memories they held into a bag.

Mr. Yee and Mr. Baek arrived and brought me outside to talk. Mr. Yee explained that a very high-ranking general would be arriving at the compound in an hour to issue Euna and me a special pardon on behalf of the chairman himself, Kim Jong Il.

"Are you happy now?" he asked, with the half grin I'd become so used to seeing.

"I can't believe it's really happening!" I beamed. "I'm finally going to see my family!"

He told me I must get ready quickly and urged me to look as presentable as possible. "Wear something bright and colorful if you have it," he said. "He's a very important general."

I looked through the clothes my family had sent me over the months. They were comfortable, casual things like sweatshirts, T-shirts, and cargo pants in muted colors. I thought back on the times when I'd receive a package and the guards would peek at the clothes curiously. They were hoping to see some pretty outfits sent from America, but they were always disappointed when they saw yet another cotton T-shirt.

"I'm a prisoner," I explained. "There's no need for me to wear anything fancy."

I chuckled to myself, thinking that my drab clothes probably helped reinforce their image of America as a poor, desolate place. I spotted a bright green collared shirt in the pile. I'd never worn it because it was too bright, the opposite of my mood during my captivity. This would have to do for the general. It was the only option.

Euna and I were brought into a small room in the compound. John Podesta and Doug Band from Clinton's team were seated along with several North Korean officials. President Clinton was not present. Two empty seats were reserved in the corner for the two of us. As the general entered the room, we all rose to our feet. He was a tall, full-framed man with a large, round face. He motioned for everyone to be seated. Through the open doorway, I saw Mr. Yee looking into the room. In all the time I'd spent with him, I never knew him to be a very expressive person. But there was something I saw in him at that moment that struck me. As the door closed, leaving him out in the hallway, I could tell he was genuinely happy for me.

It was a brief ceremony. The general spoke and Mr. Baek

translated. As at the trial, I could tell that Mr. Baek was nervous. This was perhaps the biggest, most important translating assignment he'd ever had. He was literally interpreting orders that had been handed down by the Dear Leader himself. I could see that his hands were shaking as he frantically scribbled down the general's words. He didn't want to make any mistakes.

I remembered just how crucial a role Mr. Baek had played for me during my captivity. In a country where I was handicapped by not knowing the language, he became my voice. When he translated for Mr. Yee or other officials, he never missed a beat. He was always friendly and sanguine. Just seeing him at times cheered me up.

The general announced that Chairman Kim Jong Il was pardoning us for our crimes. Hearing that statement coming from this high-ranking official was like being resuscitated from a deep coma. I breathed in deeply and looked over at John Podesta and Doug Band and wanted to embrace them and thank them for coming to our rescue.

At the end of the ceremony, Euna and I were led back to our rooms. Mr. Yee told me to sit as if I were being investigated once again, a reminder that I was not yet free. He assumed his normal position at the desk and Mr. Baek took his usual seat beside me. I was given a sheet of paper and a pen and told that I must write a letter to Kim Jong Il thanking him for his compassion.

"Dear Chairman Kim," the letter began.

If this was the final thing I had to do in order to go home, I was happy to do it. I scribbled down a couple of sentences, apologizing for my actions and thanking Kim for pardoning Euna and me.

After writing the letter, I went through some of my belongings and gathered a few items I wanted to leave with some of the people I had met. To Mr. Baek, I presented the book *The Forever War* by the *New York Times* correspondent Dexter Filkins. Iain had sent me the hardcover book, which I had started reading back in Los Angeles before my detainment. When it arrived in Pyongyang, I saw the dog-eared page where I had left off, and it made me think of my life before I was a prisoner. I imagined reading it at home just before dozing off to sleep. The book was about the rise of and battle against Islamic fundamentalism in Afghanistan and Iraq. Mr. Baek loved learning about foreign policy, so I thought it was an appropriate gift to leave him.

"This book is about the forever wars taking place in Iraq and Afghanistan," I explained. "I give it to you because I hope that this war, the war between the United States and North Korea, is one that doesn't last forever. I do hope our countries can find common ground, and that I will one day see you again."

He accepted the book graciously with his kind, warm smile.

I told Paris to distribute my toiletries and clothes among the guards and caretakers. But I wanted to leave her with something special. I took out the cashmere sweater Iain had given me for my birthday several years ago. I never wore it in captivity because by the time it reached me, the weather was already scorching hot. I also gave her the sweet-smelling shampoos Iain had sent to remind me of our vacation together in Napa Valley.

"I can't accept these things," she said earnestly. "These are special gifts from your husband."

"I want you to have them," I said. "That way you will have something to remember me by."

She accepted the items and said, "Thank you, Laura. I won't ever forget you."

To one of the female caretakers who had been trying to learn English, I left a Korean-English dictionary. She thanked me and said to me in Korean, "Now that you will see your husband, you can try to have a baby."

At one point during my captivity, while she was looking at one of my wedding photos, I had told her that I felt remorse about not trying to start a family sooner.

"I hope so!" I said to her.

Finally it was time to say good-bye to Mr. Yee. I reflected on the hours and hours each day when he would berate me for not answering his questions to his satisfaction. At first I dreaded his visits and the grilling sessions he put me through. A single look from him could make me shiver. But over time, as we opened up to each other, I began to see that he was trying to help me, that he was trying to provide me with the knowledge and information I needed to convey to my family. He was my captor and my protector.

There was a sketchbook that Iain had sent to me about a month earlier. I specifically refrained from writing in it because I was hoping I might be able to give it to Mr. Yee as a parting gift when the trial was over, but he had left so unexpectedly. It resembled the red notebook he used during the investigation. I recalled the countless times he would enter the room with the notebook and pen in hand. I'd become queasy just watching him open the red cover and flip through the pages to a blank sheet before beginning to interrogate me.

I handed him the black sketchbook. "This is a new notebook for your next investigation," I said. "Though I hope you never have to use it for those purposes. In fact, I hope you get to use it for other reasons entirely." I went on to tell him that perhaps he might even write a story about a North Korean investigator and an American prisoner whose unlikely bond becomes a metaphor for possible warming of relations between their countries. He laughed.

I could tell he was touched by this gesture. I knew I might be crossing the line when I stood up and embraced him, but I didn't care. He, more than anyone else I met, had been the most vocal with his anti-American sentiments. I wanted him to feel a connection with someone from the enemy nation. He held his body stiffly as I hugged him and thanked him for keeping his promise to get me home.

There were several cars waiting outside the compound to take us to the airport. John Podesta asked one of the North Korean officials who had been at the ceremony if he and Doug Band could ride together with us, but it wasn't allowed. I later learned that this worried Band, who couldn't rest until Euna and I were safely on the plane. Euna and I were placed in a separate car behind the others. As we drove out of the compound gates, I heard the shrieking yelps of the guard dog for the last time.

On the way to the airport, I could see Doug Band in the car in front of us looking back to make sure our vehicle was following theirs. We arrived at the airport and waited in the car while a giant motorcade of several black Mercedeses and a limousine carrying President Clinton pulled close to the private plane. The car we were in did not join the motorcade but stopped a way back from the activity taking place ahead. A group of North Korean photographers was gathered on bleachers next to the plane and snapped shots as the former president exited the limo and strode to the steps leading up to the plane. It seemed that the North Koreans, who had been orchestrating Clinton's visit from start to end, did not want to show Euna and me warmly greeting the former president on their soil. Instead, they waited until Clinton was on the aircraft before instructing us to get out of the car.

I later learned from Justin Cooper, who was in the limousine with President Clinton, that when they got to the airport, they looked around for Podesta, Band, Euna, and me but could not see us because our cars were made to wait quite a distance away from their motorcade. President Clinton and Justin Cooper were rushed onto the aircraft, where they figured we might already be, but we had still not been allowed out of our vehicles. North Korean officials then motioned for Straub, Kwon, and the U.S. Secret Service agents to board the aircraft. A moment of worry and confusion washed over the team, as they still did not know where the four of us were. It was only when they spied us getting out of the cars and headed in their direction that they felt relief.

The driver opened the trunk of the vehicle and an official motioned for us to collect our bags and go. We quickly grabbed our belongings and rushed toward the plane. Doug Band came toward us and kindly offered to help with our bags. He had stayed behind rather than entering the plane with President Clinton so that he could make sure nothing happened to Euna and me. I thanked him but declined the help, not wanting to pause for even a second. All I could think about was getting on that jet.

With each step I took up the stairs to the plane, I felt closer and closer to home. President Clinton was waiting for us at the entrance to the aircraft. He greeted us with his warm smile. I was overwhelmed with emotion when I entered the plane. No longer was someone monitoring my every move; no longer did I have to watch my words. I was no longer scared.

Everyone on the plane, including President Clinton, his staff, the pilots, and the secret service agents, was jubilant.

"I feel freedom!" I exclaimed exuberantly.

"Just wait until we're out of Pyongyang airspace," said someone on the plane, and we all laughed.

But he was right. As the plane took off, the feeling of being in the air, with Pyongyang becoming a distant speck below us, was magical. My isolation in the most isolated country in the world had finally come to an end.

·❦· **LISA**

AT AROUND 3:30 P.M. Pacific standard time, CNN began showing video of Laura and Euna walking on the tarmac toward President Clinton's plane. These were the first images that had been seen of the girls in nearly five months. We ran to the TV and replayed the video over and over again. We tried to scrutinize how my sister looked after so many months in captivity. She was wearing a green polo shirt, probably one Iain had sent, and her hair was in a ponytail. Though she looked pale, she looked healthy. Iain was the only one of us who had ever met Euna. He said she looked much thinner than she was when he met her. But no matter what, they were coming home. We all began crying tears of joy. The girls were finally free.

At 8:20 A.M. Pyongyang time on August 5, President Clinton's plane took off from North Korea, with Laura and Euna Lee inside. It was 4:20 P.M. on August 4 in Los Angeles. Laura would be home in a matter of hours. Shortly after the news reported that the plane had left Pyongyang, I received an e-mail from Doug.

"We have them," he wrote. "We're on our way to Japan. They're both doing well."

Minutes later another e-mail said, "They're in good stead, relaxing and having juice. We're all trying to be Jewish mothers."

I ran into my mother's arms, then into my father's. Mom and Dad then embraced each other, and I gave Iain a huge hug.

"She's coming home," I whispered to my sister's eager husband. "It's finally over."


∿∾ **LAURA**

O N STEVE BING'S PRIVATE PLANE, Euna and I were given a separate room with two beds. I was exhausted after having not slept in three days. But I was too wired to rest.

Our families had packed a bag for each of us that was filled with snacks for the plane ride and clean clothes to change into. My family sent goodies they knew I loved, but ironically they were Korean snacks. Even though I hadn't been given anything like this in North Korea, it seemed too awkward to eat Korean-style popcorn at a time like this. I was craving a slice of pizza.

Euna and I sat at the front of the plane along with President Clinton,

John Podesta, brothers Doug and Dr. Roger Band, and Justin Cooper. David Straub and Min Ji Kwon were sitting close by. They were all huddled around us with looks of concern. President Clinton asked us how we felt and wanted to know if we'd been treated fairly. I told him that while there had been violence when we were first apprehended, since then, we hadn't been mistreated. I was in a state of shock and disbelief that after nearly five months of being held captive, I was now looking into the eyes of the former leader of the free world and telling him our story. I wanted him to know how sorry I was that all of this had happened, but that I hoped his meeting with Kim Jong Il might have a positive impact on the state of U.S.–North Korean relations.

I told the president that we knew for certain we were going home when we saw the photo of him and Kim Jong Il on the evening news. I commented on Kim's beaming grin compared with his stoic expression.

"I had to practice that," he said, smiling. "Seriously, I wanted to be very careful not to smile or smirk. Hillary and Chelsea had to coach me."

The president also told us one of the reasons why he thought Kim was so intent on having him make the trip. He said that during their meeting, Kim told him how much he had appreciated the phone call Clinton made to him to express his condolences when his father, Kim Il Sung, passed away in July 1994. According to Clinton, Kim told him, "You were the first one to call me, even before any of my allies, and I've always remembered that."

I learned from President Clinton and his colleagues that, to their surprise, unlike other heads of state, who tend to make grand entrances with huge processions, Kim Jong Il had been accompanied by North Korean authorities who seemed very relaxed around the Dear Leader. When Kim arrived to greet President Clinton, the first thing he said, with great passion and confidence, was "I've always wanted to meet you."

President Clinton commented on how important it was to have Dr. Roger Band present because he could evaluate the North Korean leader with a medical eye. According to Dr. Band, one of Kim's arms, which appeared to be immovable, seemed to indicate the effects of having a stroke. Clinton commented on how lucid Kim appeared despite his questionable health.

Apparently, when Clinton's team first arrived, a North Korean woman official rushed up to the group and asked about a letter from President Obama that she expected the group to have. I learned that delivering a letter from Obama had been part of previous conversations having to do with visits by other potential envoys, but it was never discussed or agreed upon as part of Clinton's visit. During the former president's trip to Pyongyang, no letter was delivered and no gifts were exchanged, save for a bouquet of flowers that was presented to Clinton as a welcoming gesture.

It is hard to overstate just how unique and momentous Clinton's visit was. Unlike a typical meeting that includes the former president, where his staff carefully crafts and prepares his schedule and security detail, visiting North Korea is like entering a black hole. There are no guarantees. While the former president always travels with Secret Service protection, the seven men accompanying Clinton had to take extra special precautions. The Clinton team, the Department of State, the White House, the U.S. Secret Service, and other U.S. agencies had to manage an extraordinary set of challenges and meticulously plan for best- and worst-case scenarios in order to pull off a kind of trip that had never been made before. On the plane I could see the loads of black bags that contained various communication devices and other equipment specially chosen by the Secret Service agents for this trip. I was told that while Clinton and his team stayed in a palatial guesthouse with ornately manicured grounds, one Secret Service agent stayed on the plane along with the two pilots for the entirety of the visit so as to make sure the plane was secure.

I was grateful to the former president and his team for all they had done for Euna and me. Since their visit to the so-called Hermit Kingdom was unprecedented, they were operating in unknown territory. I couldn't imagine the enormity of the risks involved and what they had to prepare for. I appreciated hearing some of what went on, but I know there was a great deal that happened in preparation and on the ground that I didn't know about.

It was fascinating to hear President Clinton's assessment of the reclusive Kim and his account of the balancing act he and his staff had to perform during their visit. He told us about the two-page itinerary that had been planned for his team upon arrival, including visits to various monuments. They politely excused themselves from the activities, saying they were too fatigued from the traveling.

Before President Clinton met us at the hotel, he and his staff were taken to a meeting with a high-ranking official, who, yelling at the top of his lungs, went into a tirade about how terrible the United States is and how much damage President George W. Bush had done.

Later that morning, John Podesta, Doug Band, David Straub, and Min Ji Kwon went to a separate "apology ceremony," which took place in the colossal guesthouse. President Clinton did not attend this portion of the itinerary. For safety reasons, the team made sure that no person was ever alone at any given time, so Clinton's top aide Justin Cooper stayed behind with the president in his stately quarters. The ceremony involved more irate North Korean authorities bashing the United States and chastising Euna and me for our crimes. Clinton's team bit their tongues, smiled politely, and apologized for Euna's and my actions. They had one objective, and that was to get us home.

Aside from these formal bursts of outrage, the North Korean officials escorting Clinton and his crew were generally courteous and hospitable. Originally, the plan had been for only John Podesta and

Doug Band to meet Euna and me at the hotel, but when President Clinton requested to see us, the North Koreans obliged.

The Dear Leader hosted a dinner for President Clinton and his colleagues that was also attended by North Korea's nuclear negotiator Kim Kye-Gwan, among others. The elaborate fourteen-course meal included steaks, fine French wine, and an entire fish for each person. It was clear the North Koreans had gone to extreme lengths to put together the lavish affair. During the meal, Doug Band quietly stepped out to call Secretary Clinton's counselor and chief of staff, Cheryl Mills, on the satellite phone the team had brought along. It was the only communication made between one of the Americans on the ground in North Korea and the United States. He wanted to let Mills know that they were alive and well.

"This is really something!" President Clinton said to us, beaming. "You know, when I got the call from Al and the request from your families to come here, I was talking it over with Hillary and Chelsea. And Chelsea said, 'Dad, you have to go. What if it was me over there?' When Chelsea said that, I knew we had to come get you girls. I'm so happy right now." Sitting next to him at that moment, I was captivated not only by the former president's charisma and electrifying charm, which is so often talked about, but by the real concern he has as a parent and a father.

President Clinton went on to talk about a special performance that Kim wanted him and his team to attend. It was an acrobatic performance featuring ten thousand children in a stadium filled with a hundred thousand people who were awaiting their arrival. Clinton pretended not to hear the invitation, and left John Podesta and Doug Band to do the backpedaling during dinner. While Kim Kye-Gwan, who was seated next to Doug Band, continued to insist that the group watch the performance, Podesta and Band took turns yawning, politely trying to emphasize how exhausted they were from the day's events. In the end, the team did not attend the grandiose performance. It would have looked bad to the outside world to see the former president yucking it up at a performance with the leader of one of the most repressive regimes on the planet.

President Clinton also talked about some of the recommendations he gave to Kim, including releasing a group of South Korean fishermen and a South Korean businessman who were also being held by the North Koreans.

"I told him, you see what happens when these girls go home and how the international community reacts," Clinton said. "You'll get a similar reaction if you release the South Koreans."

Clinton also advised Kim to allow Stephen Bosworth, the U.S. special representative for North Korea policy, to come to North Korea to discuss how the two countries could move toward getting back to the six-party talks.

Within six months of our release, Kim Jong Il acted on all of Clinton's suggestions, and following the trip by Bosworth to Pyongyang, North Korea seemed to be making real steps toward reengaging in nuclear disarmament talks. I'd like to think that President Clinton's visit paved the way for improved relations between the United States and North Korea. Only time will tell.

·❀· LISA

M Y PHONE RANG ABOUT an hour after Doug's last e-mail. "Li," my sister said, "I'm on President Clinton's plane!"

"Oh my god, baby!" I screamed. "Did you know he was coming?"

"No, it was a surprise."

"I'm so excited to see you!" I also told her that there would be media on the ground when she arrived and that she might want to express some words of gratitude.

"Okay, I will," Laura said. "I'm going to try to get some rest. I'll call you in a few hours. Tell Iain and Mom and Dad that I love them and will see them soon. I love you, Li."

I could hear the freedom in my sister's voice.

"Oh, wait!" I interrupted. I couldn't let her off the phone without asking one more thing that I had to know. "What's President Clinton doing right now?"

Laura chuckled. "He's sitting right here with me watching me talk to you on the phone."

That made me smile.

## ∽⌇ LAURA

WHEN I HUNG UP the phone with Lisa, President Clinton told me how well she had carried herself publicly, saying all the right things to the press at the right times.

"My sister is relentless," I said. "I couldn't have gotten through this without her."

We landed at an American air force base in Japan to refuel and get some breakfast. It was our first non-Korean meal. I piled eggs, waffles, and fresh fruit onto my plate. David Straub had to advise me to go easy, that my stomach might not be used to Western food after having been deprived of it for so long.

Two flight attendants got on board with us in Japan for the remainder of the trip. It was then that I learned that when Clinton and his team were headed to North Korea to retrieve Euna and me, they stopped to refuel in Japan before heading on to Pyongyang. This was because they wanted to ensure that the plane had sufficient fuel so that they wouldn't have to purchase any gas from the North Koreans or to take any petroleum from them upon their departure. Additionally, the Secret Service agents felt it would be safer for the attendants to stay over in Japan while Clinton and his staff were in North Korea. Having them in Pyongyang would only present additional liabilities.

When we were back in the air, I tried to get some rest, but the excitement of seeing my family kept me up. Justin Cooper told me that I could feel free to use the Internet, but the thought of browsing the Web for news or checking my e-mail, something I hadn't done in five months, seemed too overwhelming. As Clinton worked on his Sudoku puzzles and dozed off at the front of the plane with his glasses perched on the tip of his nose, I decided to write the speech that I was expected to give upon our arrival. Not knowing if anything I was writing was sounding coherent, I read the speech to John Podesta and Doug Band and called Lisa to go over it with her on the phone.

At the back of the plane, I changed into some clean clothes that my family had sent along. As we made our way closer to California, a strange feeling washed over me. I found myself wanting the ride to last a little longer. Though I couldn't wait to see my family, I was also nervous about seeing how they had changed. I imagined my father and mother looking older from worry. I wondered if Iain would look at me in my pale, weakened state and still have the same loving eyes as before. After being isolated for so long, I feared the throng of media that awaited us. I had spent a career reporting on other people and issues; now Euna and I would be the ones being spotlighted. I did not want that kind of attention. I called Lisa to calm my nerves.

"Don't worry about a thing, Baby Girl," she said. "We're all here waiting for you. Mom has your favorite soup ready."

◈ **LISA**

**A** CONFERENCE CALL WAS CONVENED with several members of the White House and State Department press offices at 7:00 P.M. on August 3. It was decided that the press would be alerted around midnight that the plane was landing in Burbank. A member of the White House press office then said something that shocked me.

"We've asked President Clinton to remain on the plane while Miss Ling and Miss Lee come down to greet their families," the White House contact said.

I wasn't sure I had heard him properly. "Excuse me, did you say that President Clinton would be staying on the plane?" I asked.

"That's right, Lisa," the voice confirmed. "This is about the ladies being reunited with their families, and we'd like to have just them come down the steps."

"Uh . . . don't you think that would look really awkward if President Clinton doesn't at least come down after traveling all the way to North Korea to get the girls?"

"I'm sorry, Lisa. We feel strongly about this decision."

I was utterly dismayed by this. How would it look if President Clinton just stayed on the plane? What would people think? I was dumbfounded. I realized the sensitivities involved from a geopolitical standpoint: the U.S. government didn't want the world to think that North Korea had been rewarded by a Clinton visit. There's no question that rumors would fly from every direction about what the U.S. government would be trading in exchange for the release of the girls. I understood why the Obama administration wouldn't want to invite speculation on this, but surely President Clinton's absence upon the plane's arrival would provoke far more questions. Another conference call was convened at 11:00 P.M. and the plan hadn't changed. I brought up my concerns again, but they were shot down.

"We feel strongly about the decision for President Clinton to remain in the plane, Lisa," a voice on the conference call firmly insisted.

Just after midnight, Laura called again. She wanted to discuss her speech and to make sure to get the names of people to thank. "Don't forget Ambassador Foyer, Linda McFadyen, and Kurt Tong and the entire State Department," I reminded her, even though she had yet to meet Linda or Kurt.

She was going over her lines word for word. My baby sister had spent nearly five months in near isolation inside a totalitarian state, and now she had to give a speech that would surely be broadcast all over the world. It would be the most important speech of her life on the most important day of her life. I wished there was something I could do to calm her nerves, but my own were out of control. After going through it a few more times, Laura told me she was going to try to rest. She was suffering from delirium and exhaustion at the same time.

"Go to sleep, Baby Girl," I said. "I'll see you in the morning."

I couldn't believe what I had just said. But it was true. I would be holding my sister in my arms in less than six hours.

At some point, Kurt called to tell us to be at the Burbank hangar by 4:30 A.M. The plane was scheduled to arrive by 5:00 A.M. I got into bed excitedly, but I was still uneasy about the decision not to have President Clinton come down from the plane. I popped out of bed at about 1:30 A.M. and fired off one last e-mail to my White House and State Department contacts.

> Hi All,
>
> It's really late on the east coast but as someone who works in the media, I would be remiss if I didn't say one more time that keeping President Clinton on the plane may very well invite a whole shit-storm of speculation and chatter that you may not want. I'm fairly certain that he'd be fine with not saying anything, but to have him stay on the plane is just awkward.
>
> Obviously, the most important thing to all of us is to reunite with

*the girls, but from a public standpoint, I just think it would be so much*
*cleaner to see Pres. Clinton, and thank him.*

*In the end, it's not my call, but I feel compelled to put it out there*
*again because I'm programmed to think about possible consequences of*
*public appearances.*

*Thanks everyone for EVERYTHING!!!*

*L*

I knew that pundits would be looking for reasons to pounce on President Obama. Speculation about why President Clinton didn't come off the plane would be like feeding meat to wolves—it was just the kind of thing they were looking for. I wanted to avoid that because people had been working too tirelessly to get to this point.

We arrived at the hangar promptly at 4:30 A.M. Even at that early hour, throngs of press were lined up around the building. We were ushered into a room in the back where Al Gore, his national security adviser, Leon Fuerth, and Current's CEO, Joel Hyatt, were already waiting. We all embraced and then turned our attention to the TV that was tuned to CNN and carrying live coverage of what was happening just outside. I knew that a lot happened behind the scenes when Gore was trying to figure out a way to go to North Korea, so I thanked Fuerth for all the work he did that we weren't even aware of.

A woman from the State Department pulled me out of the room into the hallway to tell me something.

"President Clinton will, in fact, come out of the plane," she said.

I was elated and relieved. Laura and Euna would come down first and be greeted by their families. President Clinton and his team would then follow and stand behind Laura and Vice President Gore as they gave their remarks. After which, everyone would finally go home. It was a good plan.

I went back to tell my family and Al Gore that Clinton would be getting off the plane. Everyone was surprised to hear the news but thrilled nonetheless. Gore didn't say anything; he just smiled approvingly to say, *Good job.*

We were now minutes away from the moment our family had been praying for after nearly five of the longest months of our lives. We were glued to the television in a way that I imagine folks were tuned in to the first space shuttle launch. Iain was pacing back and forth in the back of the room. I walked over to him and gave him a hug. He was shaking.

When images of the shiny white 737 started to emerge, an official-looking person came to take us from the private room into the center of the enormous hangar. This is where we would stand to greet Laura and Euna as they descended from the steps of the plane. I was blown away when I looked around to see about one hundred television and photography crews roped off to the side. The camera flashes were blindingly relentless and it felt strange and uncomfortable to be on the other side of the ropes.

I told my family that Iain should stand immediately at the base of the plane with Michael and Hana. I wanted Iain's to be the first face Laura saw when she came out of the plane. My parents would be behind him. Paul, Iain's brother, Charles, and I stood behind with the Saldate family. Vice President Gore, Joel Hyatt, and Leon Fuerth hovered around our families.

∽∾ **LAURA**

**A**s we neared the Burbank airport, the sun emerged and began transmitting a brilliant golden gleam across the horizon. Suddenly my uneasiness faded away, and all I wanted was to get off the plane. The jet pulled into Steve Bing's hangar, and peering through the captain's window, I saw my family instantly. I also saw Euna's four-year-old daughter, Hana.

Once the plane was settled, a staircase was pulled up and the door was opened.

"You go first, Euna. Hana's waiting for you," I said.

Euna began shaking in anticipation.

As I followed Euna down the steps of the plane, I was overcome with excitement. Iain's smile and the glimmer in his eyes awaited me at the bottom of the stairs. I couldn't help myself, and I threw up my arms in triumph. It felt amazing.

In Iain's arms, I felt like our world was starting to rebuild itself around us, that things would be normal again. I embraced my mother and father. As I feared, I could see the toll that my absence had taken on them. My father looked grayer; my mother seemed a little frailer. Then I saw Lisa. She, like all of my family, looked tired. We were all exhausted. With tears in her eyes, she hugged me and looked me over, wanting to make sure I was okay. I could tell she was searching me over, wanting to see if the old me was still there. Lisa then took my face in her hands and wiped away the tears that were streaming down my face. She asked, "Are you ready to give your speech, Baby Girl?"

It was time. I had spoken in English during my captivity—during the interrogation and brief conversations with my guards—but I hadn't spoken very much. There were days when I never uttered a word. Now, as I began to address the public, I found it difficult to conjure up and articulate my words.

I knew there were too many people to thank, and many more I would never know because of the behind-the-scenes nature of their work, but I wanted to make sure to thank Vice President Al Gore. Even when I was in North Korea, I knew he was working tirelessly to bring us home, and after our return, I learned even more about the work he did, not only to benefit Euna and me but also to reassure and comfort our families.

Many who witnessed our homecoming told me that seeing President Clinton and Vice President Gore embrace and stand side by side on that bright August morning was an uplifting moment—a reminder, during this tumultuous period of crisis in our economy and fighting overseas, of a time when things were better. For me personally, it was a symbol of brighter things to come. I'm not sure I would be home today without their help and kindness. It was the team of Clinton-Gore, or rather, Clinton-Clinton-Gore.

-⚙️- LISA

A S SOON AS VICE PRESIDENT Gore delivered his final remarks, the press was led out of the building. Our family and the Saldates were left with President Clinton and his team and Vice President Gore and his. President Clinton was recounting some scenes from his visit with Kim Jong Il when Doug Band approached with a cell phone. He whispered something into the former president's ear and pulled him aside to take the call. President Obama was on the line. We were told that it was the first time the two presidents had spoken in quite a while.

President Clinton returned to tell us that the current president graciously thanked him for undertaking the private mission of bringing Laura and Euna home. Clinton said that he and President Obama had a "wonderful talk."

I also had a message on my BlackBerry from Secretary Clinton's deputy chief of staff Huma Abedin saying that she and the secretary had just landed in Africa for the U.S.–Sub-Saharan Africa Trade and Economic Cooperation Forum, but they had all been "working on this as a team non-stop and anxiously waiting for the good news." She went on to say, "We are so relieved and are celebrating from afar."

## ᴡᴄᴄ LAURA

AFTER WE LEFT THE HANGAR, I arrived home with my family and walked through the doorway of the house I had lived in for only four short months before my apprehension. Off in a corner was the area Iain had set up to write me letters and send packages. There were piles of photos, envelopes, a scanner, and items he hadn't yet sent. It was hard to look at these remnants of our life apart. To this day, I have not watched most of the media appearances my family made during our captivity. It is too overwhelming to see the pain they endured.

I sank into the couch in our family room. My family's eyes were all around me. They wanted to make sure I was okay, that I hadn't suffered any psychological or physical trauma. I later learned that Ambassador Foyer had noticed my bandaged head during our first meeting. He had informed the U.S. State Department, but the information was not passed on to my family. I'm glad they never knew about it during those long months of my captivity. I still experience numbness in parts of my head and face due to the violent blows.

I assured my family that I was okay, that being home was all the medicine I needed. My mother and father insisted on stuffing me with food. That was the way they showed their love. I had been craving pizza, and with the first bite of a pepperoni and cheese slice I was euphoric. My mother had carefully prepared a Chinese watercress soup in anticipation of my return. It was my favorite dish and something that always made me feel better when I fell ill. With each mouthful, I felt more and more energized.

THE RETURN OF MY SISTER to our family was the happiest day of my life. For nearly five months, a part of me was missing. At Laura and Iain's house, we all plunked down on the couches, transfixed by Laura's every movement. My sister is a California girl, and I'd never seen her skin as white as it was from being locked up inside for so many months. It was obvious that she was still in a state of shock over the events of the last few days. As she said in her speech to the world earlier that morning: "Hours ago, Euna Lee and I were prisoners in North Korea. We feared that at any moment, we would be sent to a hard labor camp. . . . Now we stand here home and free."

Despite her fatigue, she recounted some of what she had gone through while in captivity. A few times, Laura would lose her train of thought or pause for moments in silence when she recalled something that hurt her. It broke our hearts to hear of the abuse she endured when she and Euna were first captured. The details caused me to have nightmares about it for days. In some of the dreams, I became very violent with those harming my sister. But at the same time, I was heartened to learn of the caring relationships that developed between some of Laura's captors and herself.

For decades, North Koreans have been taught to regard the United States as an enemy. Laura's stories were a testament to what happens when people are able to interact with others on a human-to-human level. When one is given the opportunity to look another person in the eye, irrespective of preconceived ideas about each other, things can often change, perspectives can widen. Though Laura is my little sister, I was in awe of how she handled herself in captivity.

Our family stayed at Laura and Iain's house until well into the night. We didn't want to be away from Laura for a second, but at the same time, we wanted Laura to have some private time with Iain, who had been quiet for much of the day. He just listened intently to his wife while lovingly stroking her hair and kissing her head.

When she called me first thing in the morning and said, "Hi, Li, it's me," I had to pause for a second and pinch myself, because Laura was calling from twenty minutes away, not a world away. Tears started streaming from my eyes when she said, "Come over!" It was only 6 A.M. I rushed to her house, and later that morning my mom and I planned to take her to see our family physician, Dr. Basil. We were very concerned about the trauma from the blows she'd received to her head.

It was a great relief to hear Dr. Basil say that Laura looked like she was recovering just fine and he didn't think that any permanent damage had been done.

At the doctor's office, my mom and I stepped outside to use our phones while Laura was checked out. When I came back inside, I saw that the doctor had been briefly called away to deal with something and Laura was in the room by herself. The door was slightly ajar, and I saw my sister sitting alone against the wall. She had her head so far down that her chin was practically touching her chest. I immediately flashed back to the story she told about the officials shining a light in her face to make sure she was closing her eyes and looking down whenever she was being transported around North Korea. As I looked at her sitting like that, a sadness filled me. I went in and lifted up her chin and said, "Baby Girl, you never have to do that again."

I thought about something our God-fearing grandmother told us when we were children: "When people are in trouble, they always look down, or to the left or right. All they need to do is look up, and eventually everything will be okay."

T HE FIRST FEW MONTHS of being home were an adjustment period. It took me several weeks after my return to feel fully comfortable while speaking freely and openly. My family didn't want to leave my side, and I didn't want to be alone. Lisa came over to our house daily. She still does.

Even though many people were eager to know what happened during our time in captivity, all I wanted to do was hibernate. I was also uncomfortable that my experience had become the focus instead of the plight of the North Korean defectors who have endured so much and whose basic freedoms continue to be denied. I spent weeks not wanting to leave my house—a kind of self-imposed isolation.

In my first few days at home, I made phone calls to various volunteers around the country who had organized vigils for Euna and me. It was touching to hear why some of these people, many of them complete strangers, got involved. David Ly from Los Angeles told me, "We're Americans. We had to bring you home." Meghan Miller Jedrzejczyk from Portland explained that she had lost a daughter and felt compelled to do something: "When your sister decided to share her voice, I saw her strength and was inspired to help," she said over the phone.

I broke into tears when I first spoke with Brendan Creamer, the main vigil organizer, who had amassed a network of people on Facebook to get involved. Lisa had written about what a force Brendan had been in her letters to me. He explained that while he didn't know Lisa, he had seen her work and was a friend of hers on Facebook. "I felt I had to do something more than just say, 'Hope things get better,'" he said. Brendan and these countless other guardian angels have since become members of our extended family.

ꙮ ꙮ

A LITTLE MORE THAN two months after my return, I started feeling some slight shooting pains in my abdomen. I wondered if it might be related to my ulcer. But for some reason, these bouts didn't feel like typical flare-ups. I figured it was indigestion and popped a couple of Tums to relieve the discomfort. Then, one bright Saturday morning, it suddenly hit me.

*Could it really be possible?* I wondered.

Iain was out surfing with his buddies, but I was too curious to wait for him to come home. I rushed out to the drugstore and bought a home pregnancy test.

I paced around in our bathroom, checking my watch to see when two minutes had passed so that I could read the results. I picked up the test and looked at the small round screen. A plus sign was staring back at me. I was pregnant. Iain and I were going to have a baby. I was overwhelmed with excitement. For so long I had resisted the idea of starting a family. I felt too busy and was occupied with my own issues. Now I looked at being pregnant as an incredible blessing and a miracle. I couldn't wait to tell Iain. I grabbed my phone and sent him a text message: "When r u coming home? I have a surprise 4 u."

Iain got back about an hour later. His face was darkened from the sun and he smelled of salt water. "What's this surprise?" he said curiously.

"It's on your dresser," I replied.

Grinning, he raised one eyebrow and headed for our bedroom. I followed along, not wanting to miss his reaction. When he got to the dresser, he picked up the white plastic wand-shaped tester and brought it closer to his face to inspect it.

"Is this for real? Is this what I think it means?" I could see that tears were starting to form in his eyes.

I nodded excitedly. "We're going to have a baby," I declared.

He rushed over to kiss me and instantly began rubbing my belly. We held each other, laughing and crying all at once.

It's hard to believe that in one year, I was sentenced to twelve years of hard labor, and in the same year, the love of my life and I conceived a child. I thought back to the early days of my captivity when I both feared and hoped I might be pregnant. I had spent many sad nights anguishing over the idea that Iain and I would never have the chance to have children. Now those thoughts seem like a distant nightmare. In just a few months, we will be welcoming a new baby girl into our family. We plan on calling her Li.

# epilogue

∽⟨⟩ **LAURA**

ROM THE TIME OF my captivity and since I returned home,
I have spent a lot of time reflecting on the events of March 17,
when I made the tragic decision to step foot onto North Korean
soil. A number of factors led up to that ominous moment, which
included not only my desire to tell the most powerful story but also
the trust we placed in a guide who, at the time, seemed reliable and
cautious. But in the end, there are no excuses.

You can plan and prepare, but when real-life events unfold,
sometimes you have to go with your gut and hope you are right.
On that gloomy, frigid morning, my instincts were wrong and the
implications were enormous. For 140 days, I lived in utter terror, not
knowing if I would ever see my family again; I subjected the people
I love most to the nightmare of their lives; I caused the U.S. government to spend valuable time trying to secure our release; and I may
have put the brave North Korean defectors who shared their stories

with us in more danger than they already were. These demons haunted me throughout my captivity and continue to do so today.

Every day now I reflect on how fortunate I am to be home. I recognize the unique circumstances that contributed to our release, including having a former vice president as the chairman of the company I work for and a sister with influential connections in the media world. Many people do not have these kinds of ties. In fact, countless journalists and filmmakers remain behind bars today simply for trying to expose the truth. It is important that we keep them in our thoughts and prayers.

I hope something positive can come out of the story of my and Euna's captivity, including an increased awareness of what's happening along the Chinese–North Korean border. North Korean defectors have endured unimaginable hardships and suffering within their homeland and beyond its borders, and their stories are in grave need of attention.

I often think about the people I met during my captivity. I wonder what Mr. Yee, Mr. Baek, Min-Jin, Kyung-Hee, Paris, and the others are doing and what kind of lives they are living. Have they gone back to their old jobs? Is their patriotism for their country as fervent as it was before? Did their experience with me change their views of Americans or the United States in any way? I will always be grateful for the glimmers of compassion and humanity they showed me, and I hope one day they and their fellow North Koreans are allowed the freedom to determine their own destiny.

LISA

WHAT HAPPENED TO MY SISTER and Euna Lee makes me more eager than ever to make sure people know about the colossal humanitarian crisis inside North Korea and on its border with China. My sister's team was silenced from telling the story, but those who've risked their lives to escape dire conditions in North Korea continue to find themselves in another kind of tragic uncertainty in China. There are literally thousands of people who live in the shadows, unable to be free in China because they fear being caught and repatriated to North Korea, where they would face near certain death. Their stories are heartbreaking.

Though my sister's ordeal was the most trying of her life and our family's lives, I am proud of what Laura was doing. But just because she's home doesn't mean the story is over. I am reminded of the importance of journalists out in the world, often away from their own families, because they are determined to bring these stories to light. The only way things can change is for people to change them. But people cannot change things they don't know about. It's the job of journalists to raise awareness about what's happening in the world. And there are so many stories to tell.

Anyone interested in helping the people who've made it out of North Korea should support an organization called Liberty in North Korea, or LiNK, a terrific group that works to provide better livelihoods for refugees. They are on the front lines trying to bring attention to this human crisis. Their website is www.linkglobal.org.

Our family was also helped tremendously by two organizations that work to ensure that the rights of journalists are protected and not squelched by governments. They are the Committee to Protect Journalists (CPJ), www.cpj.org, and Reporters Without Borders (RSF), www.rsf.org. Both CPJ and RSF were our voices when we could not speak due to the sensitivities involved.

FINALLY, WE BOTH WANT to address the government of North Korea:

Should anyone from your government read this book, we want

you to know that though we have unique but independent perspectives on North Korea, neither of us ever had any malicious intentions when visiting your country. **We just firmly believe in the fundamental right of people to be free.**

We met some extraordinary individuals in your country and they treated us with kindness. For that we will be forever grateful. We truly hope that one day your people will be able to experience the kind of freedom all human beings deserve.

Sincerely,<br>Laura Ling and Lisa Ling

Just because many people have asked . . .

Mom's Special Watercress Soup Recipe

*1 whole chicken breast with bones*
*2 pork ribs*
*4 to 6 dried red dates*
*¼ cup dried almonds*
*2 to 4 small dried cuttlefish (optional)*
*2 bunches of watercress*
*¼ cup goji berries*
*Salt*

1. Blanch the chicken and pork for approximately 1 to 2 minutes.
2. Bring a pot of water (approximately 10 cups) to a boil. Place the chicken, pork ribs, red dates, dried almonds, and cuttlefish (if using) in the pot and bring to a boil again.
3. Turn the heat to low and simmer for another 3 to 4 hours.
4. Remove the bones from the stock. Add the watercress and goji berries.
5. Bring to a boil, then turn the heat down and cook for another 30 minutes. Add salt to taste.

# acknowledgments

WE HAVE SPENT COUNTLESS hours trying to think of every person who helped in the effort to bring Laura and Euna home. We've tried to come up with all of the names, but inevitably there will be some we have forgotten. For that we are deeply sorry. Know that every prayer and positive thought was incredibly helpful and meant the world to us. So thank you.

First, we must acknowledge our loving parents, Douglas Ling and Mary Ling. We opened up elements of your pasts that you probably didn't ever want exposed. We deeply appreciate your willingness to let us share parts of your life—particularly the challenging parts. We would not be the women we are if it weren't for everything you've done for us. We thank you, Dad and Mom, for your unconditional love and support.

The same goes for our husbands. Iain and Paul, we are so lucky to have such incredible partners in life.

We thank all of the members of our huge family distant and close.

To our editor, Henry Ferris, and his colleagues at William Morrow, HarperCollins: We knew from the moment we met you that we wanted to work with you. Isn't it funny how that works? Henry, we just adore you. Danny Goldstein, thanks for helping with research. And many thanks to Seale Ballenger and Brianne Halverson for expertly handling the publicity.

To Suzanne Gluck, Henry Reisch, Jon Rosen, Cary Berman, Ari Emmanuel, and all of our other friends at William Morris Endeavor: Thanks for always having our back. Suzanne, what a tiny but powerful force you are.

To Vice President Al Gore and Joel Hyatt: Thank you for providing the opportunity to raise awareness about stories that need to be told. Thank you for your relentless efforts to bring Laura and Euna home.

To the Current TV family and the Vanguard team: Thank you for your love and friendship. Your work continues to be an inspiration to many. Adam Yamaguchi: Thank you for the invaluable support you provided to our family.

To Ambassador Mats Foyer and Johan Eidman: Thank you for your unwavering persistence and your kind hearts.

To our little crisis management team of Alanna Zahn, Jeff Rose, and Morgan Wandell: Words cannot express how much we appreciate all of the time and effort you expended to help us. We love you deeply.

To the entire Chopra family: Thank you for helping to keep Laura and Euna's story in the press. Gotham and Mallika, your friendship means everything to us. Thank you for helping us with Care2, Good, Causecast, Beliefnet, Huffington Post, and AC360.

To the lauraandeuna.com team of Morgan Wandell, Andy Cheatwood, Grant Kindrick, Lisa Chudnofsky, Cindy Lin, Jennifer Fader, and Chris Mendez: Thank you for donating your valuable time to build the most beautiful website. With your help, we collected more than one hundred thousand signatures on the petition to the government of North Korea.

John and Amy Jo Gottfurcht and the team at SSI: Thank you for your constant support of Iain and our family.

Han Park, Evans Revere, David Kim, and Selig Harrison: Thank you for your concern and counsel.

Thank you to all of the friends from past and present who wrote to Laura while she was in captivity. Your words and stories provided hope and strength when she needed it the most.

Thank you, Dr. Basil and Shirley Vassantachart, and those at the Loma Linda University Medical School. We love you, Dr. Basil and Shirley. Thank you for always taking care of our family.

To Ron Olsen and Lawrence Barth of Munger, Tolles and Olsen: Thank you for your generosity, time, and wise counsel.

To Charles, Lorraine, Hannah, Jill, and the rest of the Clayton family: Thank you for being our U.K. headquarters and for sending parcels regularly to Pyongyang.

To our dear friends Jean Roh, Phil Hong, and Rick and Erin Piller: We are so grateful for the love and support you gave to Laura, Iain, and our family. Your friendship means the world to us.

To President Bill Clinton, Doug Band, John Podesta, Justin Cooper, David Straub, Min Ji Kwon, the U.S. Secret Service, and the pilots and crew who traveled so far to bring Euna and Laura home: Thank you for undertaking an unprecedented mission that was so full of uncertainty. We are eternally grateful.

Thank you, President Barack Obama, Secretary Hillary Clinton, and those in the White House and Department of State for your unwavering efforts: Kurt Tong, Linda McFadyen, Alex Arvizu, Stephen Bosworth, Kurt Campbell, Dan Cintron, Roberta Cooper, Maureen Cormack, Glyn Davies, Joseph Detrani, Daryl Hegendorfer, Sherri Holliday-Sklar, Allison Hooker, Sung Kim, Julie Kim-Johnson, Dan Larsen, John Merrill, Cheryl Mills, Jaime Oberlander, Johna Ohtagaki, Pamela Park, Amy Patel, Eric Richardson, Jennifer Roque, Ed Shin, Jim Steinberg, Jake Sullivan, Mark Tesone, the U.S.

Bureau of Consular Affairs, Janice L. Jacobs, Michele Thoren Bond, Michelle Bernier-Toth, David J. Schwartz, the U.S. Embassy Beijing, Richard L. Buangan, Linda L. Donahue, Nancy W. Leou, Bridget M. Lines, Teta M. Moehs, Maria W. Sand, Randy Townsend, William Weinstein, and the U.S. Consulate General Shenyang.

To everyone who helped with the vigils and other efforts: Thank you for taking time out of your busy schedules to help; you were lights for Laura in a very dark place. Most important, thank you for your friendship and prayers (again, if we've forgotten anyone, please forgive us): Brendan McShane Creamer, Elsa Cheung, Brenna Hamilton, Beth Diebels, Marcus Marquez and everyone at the L Wine Lounge, Elizabeth MacMillan, Jim Jordan, Isabel Johnson, Del Campo High School, Cheryl Marsh, Mike Shen, Daniel Beckmann, Holly Gibson, Kathy Zembera, San Francisco Academy of Art, Danielle Chang, Angie Wang, Rose Tibayan, Paula Rangel, Max Jones, Kim Jones, Lisa Anderson, Meghan Miller Jedrzejczyk, Nancy Loo, Teresa Cheung, Linda Yu, Richard Horgan, Stephanie Tsai, Nina May, Todd Leong, Cindy Hwang, Rob Everett, David Ly, Fred Teng, Kevin DeSoto, Dina Morishita, Welly Yang, Teddy Zee, Woody Pak, Alex Castro, Michael Beynart, Michael Fitzgerald, Joe Omar Gonzales, Kristina Velasco, Maya Lin, Lynn Perkins, Rebecca Delgado, Jeff Ong, Mindy Lee, Sara Mibo Sohn, Benson Lee, Lucie Morillion, Asia Liu, Karen Leigh, Janice Lee, Lisa Chung, Erik Gregory, Nzinga Shakur, Sean Donovan, Lisa Jenkins, Joyce Mar, Jennie Chau, Pastor Ken Joe, Trevor Debenning, LeeAnn Kim, Narcissus Allen, Stephanie Tomasegovich, Scott Ichikawa, Marcus Kwan and everyone at Wokano, Sharon Ito, George Huang, Hugh Hung, Kelly Wald, Steve Liu, Jason Martin, Hannah Song, Tia Carerre, Kelly Hu, Takoa Statham, David Kater, Chris Pham, Bobby Choy, Tom Plate, Mia Kim, Michele Chan, Susie Suh, Mark Dacasascos, Chin Han Ng, Sam Kang, Cynthia Cheng, Serena Kung, Joe Baker, Jan Yanehiro, Wendy Tronrud, John and Laurel Kao, Reverend Jesse

Jackson, Jesse Jackson Jr., Nickie Shapira, Wendy Walker, Diane Sawyer, Larry King, Margaret Aro, Jon Klein, Anderson Cooper, Charlie Moore, Bob Dietz, Tim Kelly, Chris Albert, Bombu Taiko, Yukai Daiko, Diann Kim, John Frank, the Song Family, Ken Roh, Clothilde Le Coz, Joe Hahn, David Neuman, Bill Boyd, Ron Burkle, Jean Shim, Iman Dakhil, Aude Soichet, Richard Blum, President Jimmy Carter, Congressman Ed Royce, Young Kim, Congressman Dan Lungren, Governor Arnold Schwarzenegger, Maria Shriver, Senator Dianne Feinstein, Richard Harper, Senator Barbara Boxer, Ann Norris, Supervisor Mike Antonovich, Congressman Howard Berman, Mayor Kevin Johnson, Assemblywoman Fiona Ma, Governor Bill Richardson, Janice Hartly, Senator John Kerry, Frank Jannuzi, Leon Fuerth, Pastor Jim Lee, Rabbi Abraham Cooper, Steve Bing, Andrew Liveris, Maya Soetoro-Ng, Konrad Ng, Lisa Smith, British Foreign Secretary David Miliband, British Foreign Office Minister Bill Rammell, Baroness Janet Whitaker, Baroness Caroline Cox, Lord David Alton, Glyn Ford, Ambassador Peter Hughes, Susie Joscelyne, Mayor Ray Mallon, the Korean Peninsula Desk at the British Foreign Commonwealth Office, Dr. Jeffrey Boutwell and members of the Pugwash Conference, Lucy Keung, Robert Picard, Dr. Urs Lustenberger, Robert Hathaway and others at the Woodrow Wilson International Center for Scholars, the Committee to Protect Journalists, Reporters Without Borders, the Asian American Journalists Association, and Amnesty International.

To Euna: I will never forget the courage and bravery you exhibited during those first six days of our captivity. You helped give me strength and hope. Thank you for sharing your incredible heart. I will forever regard you as a special part of my family. Love, Laura.